DARWIN THE WRITER

DARWIN
THE WRITER

BY GEORGE LEVINE

OXFORD
UNIVERSITY PRESS

OXFORD

UNIVERSITY PRESS

Great Clarendon Street, Oxford OX2 6DP

Oxford University Press is a department of the University of Oxford.
It furthers the University's objective of excellence in research, scholarship,
and education by publishing worldwide in

Oxford New York

Auckland Cape Town Dar es Salaam Hong Kong Karachi
Kuala Lumpur Madrid Melbourne Mexico City Nairobi
New Delhi Shanghai Taipei Toronto

With offices in

Argentina Austria Brazil Chile Czech Republic France Greece
Guatemala Hungary Italy Japan Poland Portugal Singapore
South Korea Switzerland Thailand Turkey Ukraine Vietnam

Oxford is a registered trade mark of Oxford University Press
in the UK and in certain other countries

Published in the United States
by Oxford University Press Inc., New York

British Library Cataloguing in Publication Data

Data available

Library of Congress Cataloging in Publication Data

Data available

Typeset by SPI Publisher Services, Pondicherry, India
Printed in Great Britain
on acid-free paper by
Clays Ltd, St Ives plc

ISBN 978-0-19-960843-0

1 3 5 7 9 10 8 6 4 2

PREFACE

I don't need to read Darwin to know what he says!

The extraordinary Darwin bicentenary year of 2009 kept virtually everyone who has ever written a word about Darwin very busy. For me, it was an annus mirabilis, taking me to lots of places I didn't expect to go, finding everywhere people excited about Darwin and evolutionary biology and the related controversies, and rushing to museum exhibits, lectures, and conferences. The year left me so filled with Darwin thinking and Darwin lore that I realized that I couldn't let it all just dissolve into 2010 when, I expected, there wouldn't be a lot more invitations to talk about Darwin. The experience of living and breathing Darwinian ideas, reviewing yet again Darwin's life, debating again the apparently infinitely extending Darwinian conflicts, and re-reading Darwin's own language, did nothing to diminish my enthusiasm for his work, and for his life. I have emerged from the bicentenary celebrations with about six Darwin-related t-shirts, a Darwin sweat shirt, several Darwin dolls and pens, three Darwin caps, and a Darwin bumper sticker. I would happily find more.

I write, then, from a perspective not quite that of the "objective" and distanced scholar, whose commitment to knowledge must short-circuit the enthusiasm that in almost all cases scholars feel for subjects they have studied intensely. I have studied Darwin intensely; I have registered his flaws, his limits, and some of the dark consequences of his extraordinary vision. I have tried to follow his arguments and the development of his arguments in later evolutionary science and into the evolution battles that emerge from school boards across America. I have done my best—probably not good enough—to keep up with the production of the enormous Darwin industry that has helped make him perhaps the most documented figure in world history. And from it all I have emerged writing as an unrepentant partisan both of his ideas and of his art. In this book, I want to focus on his art, although I recognize that Darwin would probably have been surprised to hear people say he had any. But while Darwin's art and writing will be front and center, reverberations from the controversies and from the extraordinary modern developments in evolutionary thinking are likely to be heard. I write neither as scientist nor as a philosopher, but as a student of literature.

As a student of literature, I am allowing myself to indulge my enthusiasm, to underplay many of the reservations that I might have about Darwin the man, about Darwin the thinker, about Darwin on matters of race and gender. My literary enthusiasms have more than survived my reservations. (Some of the process of that survival is laid out or implied in my earlier book, *Darwin Loves You*.) Although I really like the Darwin I have met in hundreds of his letters, in his writing, in the many biographies, and in his notebooks, I know that he was no saint and deserves no hagiography. I distrust hagiography anyway. Many of his flaws (*relatively* few and *relatively* normal, however) have been intimated or dramatized in Adrian Desmond and James Moore's excellent biography, *Charles Darwin: The Life of a Tormented Evolutionist*. Some of Darwin's ideas have been taken into awful places; he was himself much impressed by the work of his cousin, Francis Galton, who moved evolutionary thought towards eugenics. Certainly much bad stuff has been perpetrated in his name. His wonderful modesty (I believe it was real) often disguised his undoubtedly self-interested moves to elicit help from correspondents. There are even passages in *The Descent of Man* that make me cringe. Finally, it would be simple obtuseness not to recognize that for many people Darwin's imagination of the world is dispiriting or even tragic.

But that is not how it has been and is for me. Darwin makes me feel better, not worse. He fills my world with wonder, turns the most ordinary objects in nature into wonderlands. And I believe strongly that there is a serious need these days for the perspective on Darwin that allows us to experience him in this way. First of all, it matters that Darwin is readable. How many who throw around his name and invoke him in the very many negative ways he has been invoked have actually read him? The subtitle of this chapter captures all too well Darwin's position in our culture. Everyone "knows" what he said. Why bother to read him? Some people really know a lot about what he said without reading him (though those people, I'm willing to bet, are few). Some people really have a sense of what "natural selection" is, how it works, what effects it has on the world we inhabit. Some people have an idea of what "evolution" means beyond the idea that we are descended from apes. But reading him gets us beyond the simple formulas that, for most people, represent Darwin's ideas—it's a dog eat dog world, it's survival of the fittest, it's every man for himself, it's a depressingly mindless world. Actually reading him can change fundamentally not only the way we think about him and his books but the way we look at the world and the processes of nature that are everywhere around us (and in us).

One of the points of this book is that "actually reading" something is considerably different from "knowing" what it says. In fact, actually reading something changes—and certainly it does in Darwin's case—what a book can be taken to mean. The words matter. The metaphors matter. The construction of arguments matters. The "affect" of the words matters. The implicit attitude toward what is being said matters. Knowing what a book "means," knowing what a book "says" without reading it is like having the power to detect what a mathematician feels about the world and even his own ideas by understanding his non-verbal formulae. It is like knowing what a mathematician "says" without knowing the formulae.

The experience of reading Darwin, with as open a mind as possible, is what I am enthusiastically endorsing with this book. I write with enthusiasm because "enthusiasm" is one of the critical effects of Darwin's writing. Not everyone will find the word-by-word and page-by-page reading of Darwin's famous books as exhilarating and liberating as I have found it. There will inevitably be moments of tedium, when, persisting in getting it right, and spelling it all out, he dives as deeply into the phenomena as possible, and struggles, sometimes even awkwardly, to find a precise way to phrase it. Sometimes out of that passion to be precise, to argue as convincingly as possible, an almost lyric passage will emerge. Sometimes the very dogged precision of fact and argument produces stunning effects. Sometimes not. But he was a very good writer, if not a writer like the great novelists and poets of his time. I do not want to claim that his greatness as a writer is like that of Dickens or Tennyson or George Eliot. But the nature and quality of his writing matters, and around the corner from the occasional struggle there emerges the exhilaration, and a vision of the world that makes everything in it—even the stuff we really don't like—fascinating.

Thinking and feeling this way about Darwin is obviously not the norm of our culture, and it's partly for this reason that I have felt impelled to keep writing about him, to keep urging not only the importance of what he said—even his worst enemies agree to that—but the remarkable satisfactions that come from the way he said it. Coming to terms with Darwin's world and Darwin's way of writing—I want finally to insist—is a way to come to terms with our own world. I have to premise, if it's not clear already, that however far modern science has taken us from Darwin's exposition of the theory of "descent by modification through natural selection," and however wrong he was about various often important details, he was right. What he told his world about our world was essentially true, and widespread modern resistance to those truths is destructive in truly dangerous ways. One of the ways to come to terms with Darwin's understanding of the world is to experience the language in which he first formulated it, and bring to that encounter an openness that will allow us to transcend the abstractions and simplifications of his idea that come from "knowing what he said" without reading him.

Two of my experiences relating to Darwin may help explain the urgency I feel about all of this, and the reason I am willing to indulge my enthusiasm. I have learned by telling these anecdotes many times to friends and acquaintances that these experiences are not singular. The people I encountered expressed views that are certainly representative of the views of a great many people, almost all of whom "know" what Darwin said without reading him.

First: at least 25 years ago, I was invited by the English Department of the University of Texas to give a lecture related to my current work, which was at that point about the relation of Darwin's ideas and writing to that of other Victorians, like Dickens, or George Eliot. It was a typical English Department event, with an appropriate turnout of graduate students, the usual smattering of faculty, and a few non-academic types who seemed interested in the announced subject. The talk, it seemed to me, went well, as I tried to show how some of the things Dickens did in his novels had an interesting and somewhat complicated relationship to basic ideas that lay behind Darwin's work—things as rarefied to non-academic ears as uniformitarianism and catastrophism.

I wasn't making a case about Darwin's work or even attempting to explain his famous theory. I'm not sure I even mentioned "natural selection." The name "Darwin" was in the title of the talk, but it really was about literary themes and forms. All I did was indicate that behind Darwin's ideas there were a lot of ideas shared culture-wide and that these probably were at work in novels as well as in developing science. Both the novelists and Darwin were under the influence of ideas that helped make possible the idea of evolution by natural selection. I can't say it was a thrilling talk, but for an academic audience it did its job.

After the session was over, several graduate students came up to speak to me, as is the norm for visiting lecturers, and I was enjoying the discussion when suddenly all the students but one disappeared. And that one turned out not to be a student, but someone who had come to tell me, "You know, Darwin repented on his deathbed."

I was taken aback by the comment, for which I was (naively, I now understand) entirely unprepared. If the name "Darwin" appears in a talk, no matter in what context, there's an excellent chance that listeners with anti-evolution axes to grind will turn up. It didn't matter that I knew what she said wasn't true. I had read a lot about Darwin and knew a lot about his life (and death). I had rummaged extensively in biographies and in his correspondence and notes. In fact, I knew that whatever Darwin's feelings about religion were, however humble he was about his work and his character, he never spoke or behaved or wrote as though he had in that respect something to "repent" for. So, off guard, I mumbled something like, "No, I don't think so." At which point the young (or middle-aged) lady began talking to me about evolution, but more broadly and knowledgeably about salvation.

It became simply a *tête-à-tête*, or *tête contre tête*, in which I found out a lot about what fundamentalist Christians really believed.[1] My new acquaintance told me things about Darwin and evolution that took my breath away in the extravagance of her confident misunderstanding. As the talk turned more directly to religion and we discussed the moral and spiritual consequences of not believing New Testament teachings literally, I finally asked, "Well, you know I'm Jewish. What's going to happen to me?" It was as if I had set myself up to be straight man, and her answer was immediate—if in tones without malice: "You're going to hell."

It was a straightforward and factual announcement. That, perhaps unfortunately, is the way things were. As I had the facts of literary history at my hand, she had the facts of salvation. Indeed, she was more confident about her facts than I was about mine. Hers weren't contestable, even if it meant consigning this nice chatty but alas Jewish professor to hell. I could have been saved if I had wanted to be. After all, even Darwin had been saved through repentance, though most professors don't seem to know that.

I can't say that after damnation I was particularly well inclined to fundamentalist opponents of evolution, though in fact by the end of the conversation, as I learned that this woman's husband had died young and that her child was autistic, I felt a lot of sympathy for her. And the sad facts of her personal life allowed me to build a story that at least almost excused both her religious certainty and her mistaken knowledge about Darwin. For a long time, as I reflected on this singular experience, I found myself generalizing it so as to understand better why so many Americans have decided that the Virgin birth is more probable than the evolution of organisms. People need to believe this. Life is too rough and unfair not to believe that there is someplace, some-where, that at some time will redeem the injustices, losses, and pains that mark everyone's life, but some lives, for no obvious earthly reason, a lot more than others.

So while there is a lot more at stake than the details of what Darwin literally said, I couldn't help thinking also that the exper-ience of reading Darwin might have affected in some way the totally ignorant but movingly urgent ideas of the woman who had, perhaps even reluctantly, consigned me to hell. It wasn't that she didn't like me. I understood that no argument of mine about the evidence for evolution could dissuade her, and that no argument of mine about how interesting the world becomes if you see it through Darwin's eyes could ease the pain of her personal losses, or the satisfaction of her imagination of salvation. And yet it did seem to me that reading Darwin instead of taking him as a kind of emblem might have helped, at least a little.

Philosophically or scientifically speaking, what difference would it make to the idea of evolution if its major propounder, dying, had renounced it? The implicit assumption is that once Darwin re-pents, all the evolutionary science of the past 150 years is also invalidated. Does that mean that all practicing scientists who work with the idea of evolution will also have to repent, or does Darwin's repentance mean that they are all working on a huge mistake? Science has come a long way since Darwin died. But if Darwin matters so much to deep believers, wouldn't it be important first to read what he had to say; to understand what his arguments were and to feel the impact of his language and his ideas?

Second: many years after my experience of damnation in Texas, I was having a terrific cup of espresso in a pasticceria on Bleecker Street. I was carrying a book on Darwin (I'm afraid I'm always carrying a book on Darwin), and as I was checking out, one of the men behind the counter noticed the book and the picture of Darwin on the cover. "Do you really believe in Darwin?" he asked, as he took my check. How does one answer that? He clearly didn't mean whether I believed Darwin existed, as he might have asked whether I believed God existed. "Darwin" simply equals "evolution" and knowing that, while not being very quick on my feet, I managed only to mumble something like, "Yes, of course." "It's crazy," he replied, with what I took as a look of sardonic condescencion: "Do you really think we're descended from monkeys?" Waiting for my change, I asked myself how I could handle that question. He was obviously assuming, not having read Darwin, that Darwin said we are descended from monkeys. The real test of a "believer" in Darwin would be willingness to accept the monkey as some great-great-grandfather, a concept just too absurd for the counter-man to entertain for an instant. Would it be an adequate response to begin explaining that Darwin didn't say we are descended from monkeys, but that . . . etc., etc.? Could I begin, with several customers behind me, to explain the whole Darwinian argument about branching descent? Humbled and silent, I took my change and began to walk off, feeling very inadequate to the moment. He, meanwhile, gave a knowing look at a fellow worker, who was boxing four cannoli for another customer, and called triumphantly after me: "So how come there are still monkeys?"

It was very funny, I had to admit to myself, though I was too abashed even to smile. Only some time after did I learn that this question too was not unusual at all but part of a quite common anti-Darwinian bag of tricks. "It is unfortunately necessary," says Richard Dawkins, to explain, again and again, that modern species don't evolve into other modern species, they just share ancestors: they are cousins. This, as we shall see, is also the answer to that disquietingly common plaint, "If humans have evolved from chimpanzees, how come there are still chimpanzees around?"[2] Here was no fundamentalist condemnation to hell, only a strong conviction that reason was on his side—even against the whole field of biology. What it had in common with my earlier experience in Texas was the ignorance of the dogmatically confident objector to evolution. He too had not read Darwin.

I offer these anecdotes simply as examples of the cultural condition that has helped provoke me to write this book. The book, however, is not designed as a polemic, or as another entry into the fruitless and passionate debate at the public level (among scientists it doesn't seriously exist) about whether there is such a thing as evolution, about whether there is a God, about all those Big Questions. Rather, I want to reiterate, this is about Darwin's art, and it is impelled by a sense that people ought to *read* his books; preferably before they start talking about what he said. I want to avoid polemics, discussions about why evolution is true, or about whether Darwin repented, or about why there are still monkeys. Rather, I want to turn back to Darwin's books and Darwin's language. All of the big questions have been addressed many times; the battles continue to be fought. But I want to respond to the fact that people don't read Darwin before they talk about him, they don't experience Darwin, they don't think about him as a writer but as a negative icon of some "faith," or a hero of science. There is nothing I can say to convince the woman in Texas that Darwin didn't repent (and that if he did, it didn't matter to the question of evolution). I don't even want to convince the counterman that the fact that there are monkeys doesn't confute Darwin. I want to induce them to turn a few pages of *The Voyage of the Beagle*, or *The Descent of Man*, or *The Expression of Emotions*, or *The Formation of Vegetable Mould through the Action of Worms with Observations on Their Habits* (now how can anybody resist that title?). Of course, I would like to tempt them to read the masterpiece of them all, *On the Origin of Species*. And I would like them to try to read those books just for the fun of it, just for the pleasure of reading, learning, paying attention, without their ideological armor on. That hope too is likely to be in vain. But persuading anybody just to read Darwin, as though he were a writer of general interest, is my more realistic objective here. My argument, simply put, is that read that way, Darwin, in spite of history, in spite of theological battles, in spite even of himself, can provide remarkable pleasures to anyone.

Darwin's is a different kind of discourse from the popular ones to which we have become accustomed; he's not a novelist or poet or even, intentionally, a popularizer of science, like Stephen Jay Gould in our own time. He's a scientist in love with his subject, excited about it, and he felt the necessity of being as precise and persuasive as possible, so he attended with great care to his prose, and even in his deepest determination to remain objective and cool and "scientific," he managed through that careful prose to convey something of what it feels like to see the world as he saw it. That is the work of a writer, and it is to that writer that I devote this book. Perhaps, if my confident counter-man had read Darwin, he would have learned not only that Darwin did not claim that we are descendants of monkeys, but he would have felt something of what it means to live in Darwin's world.

The first shock of my first serious reading of *On the Origin of Species* was the discovery that it was written in the voice of a man who allowed himself to express the feelings nature aroused in him, and who sought the richest metaphorical possibilities in his attempt to describe afresh what the world is like. And when the dust of celebration cleared from 2009, I found that throughout the year, it had been the language that had pulled me forward, and it was to the language I wanted to return at least one more time. This book is about how Darwin wrote, and about how what he wrote not only affected the scientific world to which he belonged, but many Victorian writers whom we value.

Everyone who dares to write about Darwin's work as literature owes an enormous debt to Gillian Beer, whose *Darwin's Plots* of 1983 opened up Darwin's work for literary scholars in a way that continues to this moment to resonate and to influence. After I had written much of this book, I discovered that a third edition of *Darwin's Plots* was being published by Cambridge University Press, and where I have been able to do it, I have sneaked in an allusion or two to it. It is still the indispensable, and by far the best, book on Darwin as a writer. As I have proceeded through the years since I first met Gillian Beer and first began writing on Darwin (that is, perhaps a year before she invited me as a fellow to Girton College, Cambridge, *and* her great book was first published), I have come to know and value the work of Robert J. Richards, the historian of science, who seems to me to have done perhaps the most remarkable and original work in getting us close to the spirit and ideas out of which Darwin's writing emerged. I could not have written this book without their work, although they have no responsibility whatever for what I have done trying to follow in their wake.

I should say here also one word (or two) about Jonathan Smith, a scholar I have known since his days in graduate school. His recent book, *Charles Darwin and Victorian Visual Culture* is, along with the work of Beer and Richards, the very best study of Darwin as a cultural figure as well as scientist. Smith, moreover, has overseen a little piece of what is now this book, and his broad and imaginative knowledge of Darwin and of the problems of science writing as cultural writing has been very important to me in my thinking about virtually every subject I raise here.

Finally, I owe debts all over the place, as far flung as the various Darwin conferences and festivals I have had the great good fortune to attend during the annus mirabilis of Darwin's bicentenary year.

To Alan Kamil, the wonderful student of bird intelligence at the University of Nebraska, who introduced me to my only Clark's Nutcracker of 2009, and a very bright young bird it was; to David Sloan Wilson, that dogged, passionate, and brilliant biologist and proselytizer for the study of evolution across disciplines in every corner of our education, and to Ann Clark, who the evening of one of my talks at Binghamton told me more about the intelligence and behavior of crows than I could have imagined or even thanked her for; to Robert Siegel and the other wonderful biologists and scholars at Stanford who were involved in the course on Darwin's Legacy; to Rebecca Stott, whose beautiful book, *Darwin and the Barnacles*, seems to have been only a prelude to further thinking and writing about science and evolution and literature; to Valerie Purton, at Anglia Ruskin University, who brought me back to Cambridge to celebrate both Darwin and Tennyson; to Suzy Anger, who invariably helps me through areas of Victorian culture I should know more about; to Keith Wilson, who read and helped me with my Hardy chapter; to Paolo Costa, a brilliant and learned philosopher/theorist, who writes English brilliantly and translates me into Italian so that I think I know the language well; to Bruno D'Udine, whose work with Aboca in Italia has been intellectually generous and encouraging; to Aeron Hunt of the University of New Mexico; to Nalini Bushnan of Smith College; to Matt Stanley at NYU, and Jim Endersby, from whom I have learned much and whom I finally met at the conference Matt organized. As almost always, I owe more than I can say in English to the Bogliasco Foundation and the wonderful support staff at il Centro Studio, particularly Ivana Folle, and in addition to the amazing group of fellows with whom I worked and talked and drank and ate, perhaps too much and too well: Marge Levine, of course, and Kathryn Davis, Eric Zencey, Justin Kramon, Orhan Mehmed, Robert and Faida Byrd, and Agatha Simon, who managed to teach me some French along the way. Grazie mille a tutti. Perhaps, however, I owe my greatest debt of all to Darwin himself, who helped change the way we all look at the world, and has made it in the process, more brilliant, more beautiful, more interesting, and perhaps a bit more frightening, as well.

Notes

1. I learned then that the myth of Darwin repenting on his deathbed is part of the fundamentalist repertoire. To respond to this persistent myth James Moore, the co-author of a major biography of Darwin, studied with great respect for those who believe the myth, and with extraordinary scholarly seriousness the history of the myth. Moore does not polemicize, but produces a careful analysis of the evidence, far more careful, I must say, than the myth seems to deserve. Aside from untrustworthy hearsay, there is absolutely no evidence to support the idea. See James Moore, *The Darwin Legend* (Grand Rapids: Baker Books, 1994).

2. Richard Dawkins, *The Greatest Show on Earth: The Evidence for Evolution* (New York: Free Press, 2009), 408–13.

TABLE OF CONTENTS

LIST OF ILLUSTRATIONS

I

Darwin the Writer

We don't read Darwin because what he says is what scientists now believe—much of it isn't. We read him because a book of eloquent argument and well-ordered evidence, assembled with such modest yet personal passion, reminds us of the powers of the human mind to bring light to darkness, make a clearing in the wood of confusion.... It is not blind belief in Darwin's view of nature but our love for what he did to the blind nature of belief that makes biology, and us, Darwinist.

Adam Gopnik

If one were to try to identify the most important book in English literature written in the nineteenth century, it wouldn't be a novel of Dickens, neither *David Copperfield* nor *Bleak House*. It wouldn't be what is probably the greatest English novel of the nineteenth century, George Eliot's *Middlemarch*. Nor would it be Wordsworth's great poem, "The Prelude." It wouldn't be "literature" in the conventional sense, at all. Rather, it would be Charles Darwin's *On the Origin of Species*, published on November 24, 1859. While by Dickensian standards it was hardly a best seller, it did sell out its first edition (with a second printing) of 4,200 copies, and went through five more editions in Darwin's lifetime.

The notorious, anonymously published *Vestiges of the Natural History of Creation* (1844), by Robert Chambers, consistently

outsold the *Origin*,[1] but the spreading impact and much higher
"respectability" of the *Origin* gave it enormous cultural weight.
Although we know, not only because of the sensational success of
Vestiges, that the idea of "evolution" (a word that Darwin *only*
introduced in the sixth, 1872, edition of *Origin*) had been broached
by many writers and thinkers long before 1859, not least by Darwin's grandfather, Erasmus, it was *On the Origin of Species* that
made the idea credible. It is *On the Origin of Species* that significantly changed the way everyone, not only the great writers of the
time, could look at the world. And although it is normally and
correctly assumed that that's because the idea was so powerful,
I believe that much of its great influence followed from its art,
that is, from the particular way that Darwin found to make his
case; the particular way the book was written.

His way of writing, I will be arguing, not only empowers and as
it were creates the ideas; it makes the book something more than
its ideas. Gillian Beer, in the study that has set the standard for
treatment of Darwin as a writer, notes that Darwin's language is
not distinct from the ideas they express but intrinsic to them; it
cannot be skimmed off.[2] Alive with metaphor, brilliantly detailed
descriptions, twists, and hesitations, and personal exclamations,
the prose is saturated with aesthetic, intellectual, and ethical
energy, and with the sorts of tensions, ambivalences, and feelings
characteristic of great literature. Many years before Beer, Stanley
Edgar Hyman, looking at Darwin as an "imaginative writer," noted
the "prophetic quality" of the *Origin*, and its predominant tone of
personal testimony.[3] The personal testimony, a striking and for
most new readers in our times an unexpected quality of such a
famously "scientific" text, is also part of that strange and delightful
domesticating of the vast and "mythic" that gives the *Origin* its
distinctive and yet extremely Victorian character. It also makes the
generalizable thought, the thought that has been central to the
science of evolutionary biology, very particular as well, and intimates yet another kind of knowledge.[4]

Darwin was as assiduous in finding the right language for his arguments as he was in the research that produced them. Early on, trying to put the notes and the diary he wrote during his five-year journey on the *Beagle* into clear prose, he wrote to Caroline, his sister:

I am just now beginning to discover the difficulty of expressing one's ideas on paper. As long as it consists solely of description it is pretty easy; but where reasoning comes into play, to make a proper connection, a clearness & a moderate fluency, is to me . . . a difficulty of which I had no idea.

In the course of his career, Darwin several times made clear that writing was not for him a natural gift. He wrote to A. C. Bates, for example,

Some are born with a power of good writing, like Wallace; others like myself & Lyell have to labour very hard & slowly at every sentence. I find it very good plan, when I cannot get a difficult discussion to please me, to fancy that some one comes into the room, & asks me what I am doing; & then try at once & explain to the imaginary person what it is all about.—I have done this for one paragraph to myself several times; & sometimes to M^{rs}. Darwin, till I see how the subject ought to go.—It is, I think, good to read one's M.S. aloud.— But style to me is a great difficulty; yet some good judges think I have succeeded, & I say this to encourage you.[5]

In a throwaway line at the end of a letter to Lyell in 1867, he wrote: "A naturalist's life wd be a happy one, if he had only to observe & never to write.—" (*CCD*, vol. 15, 1 June 1867, 286). Against his own instincts, but with the same assiduity with which he observed nature, he became a real writer because he (correctly) believed he had to.

I want, then, to consider some ways in which, although it was certainly a work of "science," the *Origin* was equally a work of "literature," not only because its cultural consequences are still being felt, or because it significantly affected how writers could imagine their fictional worlds, but because, as Adam Gopnik insists, it is worth reading regardless of whether Darwin was right about evolution and natural selection (which, by good fortune and some genius, he most certainly was). Just as readers return again and again to Jane Austen or to Dickens, one should return to (or, at least, visit for the first time) Darwin, who not only produces the famous theory for which he is best known, but reads life in its particularities so astutely as to enrich the reader's experience of it, regardless of theory. Not every one will agree that the *Origin* is a good read.[6] It does, indeed, have its gray moments. I don't want to suggest that it has the texture of the great imaginative prose writers; certainly, it aims to be and is a "scientific" text and requires of the reader an attention different from that which we would normally give to characters in fiction. With the right kind of attention, however, one becomes aware of a central "plot"—the revelation of a world, in detail after detail, habitat after habitat, in which there unfolds a complex story of death and transformation. If everyone's experience is anything like mine, reading the *Origin* is exhilarating, an eye-opener; as Gopnik again put it, the *Origin* is "a book that makes the whole world vibrate." Reading Darwin provides a remarkable, refreshing, and energizing way not only to think about evolution but to feel about oneself and about the world. [7]

I am a long way from the first critic to find Darwin's work "literary."[8] As I proceed, I will have occasion to refer to many books and essays that have found Darwin's work surprisingly charming, surprisingly metaphorical, and surprisingly full of cultural implications well beyond the quite literal arguments about the nature of nature that Darwin was trying to make.[9] I might add "surprisingly pleasurable" to that list. But I want, in particular, to emphasize another direction that Darwin's literariness takes; one that moves away from the normally assumed bleak and tragic vision that, to be sure, one certainly can also find in his prose. *On the Origin of Species* surprises most first readers with its charm, with the presence in it of a personality—its' "personal testimony"—with its excitement about the world it describes, with its wonderfully intricate but unassuming precision and complexity of argument, and with its imaginative daring. These and other qualities are not exactly what one might have expected from a "dry" scientific tract, and do not seem coherent with a disenchanted vision of the world, or with the sorts of disasters that admirers of Victorian fiction find played out in the consistently catastrophic famous narratives of Thomas Hardy, say, or the even more disastrous ones of the French naturalists. I am interested in particular in this incoherence of feeling and form with the predominant literal meaning of the book, and in exploring ways in which Darwin evokes through his prose another order of truths, and feelings like Gopnik's and my own; and I want, as this book unfolds, to suggest that those kinds of feelings are not peculiar to a few twenty-first century intellectuals, but were experienced by many writers traditionally not seen as particularly Darwinian (because they don't seem disenchanted), and by many other "Darwinian" writers (who are normally taken to be rather gloomy about the world).

I will be arguing then that Darwin's enormous cultural success—in the short run, at least—depended as much on the form of his argument and the nature of his language, as it did on the power of his ideas and his evidence. I don't want to minimize these latter; the quality of the writing itself depends on the nature of the evidence and the way it is deployed. The qualities requisite for good science—not least, precision—are not finally much different from the qualities required for good literature. But Darwin's facts often emerge on his pages already saturated with feeling— often the feeling of awe, frequently the feelings of compassion or sympathy, and sometimes even revulsion. Behind the writing of the *Origin* there is something still, as David J. Depew puts it, of the "Romantic zeitgeist" that Darwin had "imbibed from the youth culture of his day."[10] The Romanticism is most evident in the *Journal of Researches*, where Darwin could be—and was—a nature writer rather than an exponent of a new theory. Darwin experienced the exotic landscapes he encountered on the voyage in what he called, in a letter written before he knew he was going on such a voyage, "a tropical glow."[11] On the voyage, he was, as David Kohn describes him, a more or less self-conscious child of Romantic poetry and Romantic painting, of Milton seen through romantic eyes, of Wordsworth, Shelley and Lyell and German Romantic painters, like Carl Wilhelm Holdermann and Joachim Mortiz Rugendas. Like Alexander von Humboldt, Kohn notes, Darwin "self-consciously incorporated affect and imagination into his early science."[12]

"But a case can be made," DePew claims, "that the *Origin* is just as expressive as the *Journal of Researches*" (p. 238). That Darwin was a Romantic, scholars like Gillian Beer and Robert Richards, had already made abundantly clear;[13] that the Romanticism inflects the science, plays off not against it but within it, is one of the critical facts of Darwin's science, of his art, and of his writerly virtues. How the Romanticism works with the science would seem to be a problem, but it was not at all. Considering that question, Richard Holmes puts it succinctly and correctly, in a way that will matter significantly for my own arguments: although "Romanticism as a cultural force is generally regarded as intensely hostile to science, its ideal of subjectivity eternally opposed to that of scientific objectivity... the notion of wonder seems to be something that once united them."[14]

This certainly holds true in Darwin's work, where the voice of the scientist is not yet fully disguised in the uniform of professional distance that working scientists must wear today. Much of Darwin's science seems to be generated by something extra-scientific, by his pre-*Beagle* passion for the tropics, for example, but also by his naturalist's enthusiasm for bugs, birds, dogs, and ants, and worms, and by the awe he feels as he observes the abundance and complexity and ultimately sublime energy of nature in its many amazing particulars. It is the young Darwin who, completing his *Journal of Researches* and his five-year voyage around the world, speaks of his experiences of the "sublimity" of "the primeval forests undefaced by the hand of man," and it matters not whether they are places "where the powers of Life are predominant, or those...where the powers of Death prevail": "no one can stand in these solitudes unmoved, and not feel that there is more in man than the mere breath of his body."[15]

Of course this was written before Darwin became the "agnostic" of his later years, and I invoke this passage not to suggest that, yes, after all, Darwin was a religious man. In his *Autobiography*, he alludes to this passage but goes on to say that "now the grandest scenes would not cause any such convictions and feelings." But Darwin's writing, through all those remarkable books in which he sought (and found) evidence for his theory, regularly register feelings akin to that experience of the sublime he experienced on the *Beagle* voyage; no need for the "convictions" in order for Darwin to feel the wonder of the natural world he so carefully and scientifically described. The "feelings" are secularized, but they are there: "The state of mind which grand scenes formerly excited in me, and which was intimately connected with a belief in God, did not essentially differ from that which is called the sense of sublimity."[16] What needs emphasis here, however, is not the question of religion or secularity: it is that Darwin's relation to nature, at every stage of his life, was romantically intense and very deeply and personally felt. If it is not divine, it is sublime, and evokes those feelings of awe and wonder that sublimity always evokes. The awe he describes in *The Journal of Researches* came to him throughout his life if in different, more secular forms, whether the phenomena he observes are on the side of "Life," or of "Death." The sublime, at least in particular Darwinian versions, echoes through virtually all of his writing; and even before he had self-consciously moved away from Christianity, he invested the natural world with an intrinsic and awesomely overwhelming value. Darwin may have lamented his loss of feeling for poetry and Shakespeare towards the end of his life, but he never lost his feelings for the extraordinary particularities of nature as he observed them, not even for worms! That awe, stirring his never-satisfied curiosity, provoked him to his last book, the improbably—for a best-seller—entitled, *The Formation of Vegetable Mould through the Action of Worms with Observations on Their Habits.* "Worms," Darwin claims, "have played a more important part in the history of the world than most persons would first suppose."[17] How can one resist a sentence like that? And that kind of intensity—huge claims wrapped in understated prose; great claims analyzed into little ordinary details—never left Darwin, and is distinguishable even in the much less personal writing of *On the Origin of Species.* It does not there, or anywhere else in Darwin's post-*Origin* work, point to some transcendental reality, but it does register the excitement and passion of his engagement with nature, in all of its (sometimes deadly) phases.

It is no wonder then that as he contemplates and imagines Nature, Darwin works almost instinctively in metaphors, which do a lot of very scientific as well as rhetorical work—beginning with "natural selection" itself. Notoriously, this central metaphor, whose literal meaning has been taken to signify a Nature utterly indifferent and utterly mindless, embodies a scrutinizing, caring, very maternal (and certainly not diabolically cruel) figure. As Robert Richards shows, in Darwin's early drafts of his theory, he "worked out for himself the character of natural selection; and that character was cast in the image of a divine Being, whose 'forethought' might teleologically produce creatures of great 'beauty' and with progressively intricate 'adaptations.' Natural selection, in its original, metaphorical conception, was hardly machinelike, rather godlike."[18] Daniel Dennett calls natural selection an "algorithm," a mere recipe for the processes of nature, but this is certainly *not* the "natural selection" of *On the Origin of Species.* It is, however, a perfectly reasonable term for the processes, but only if one extracts "natural selection" from the language Darwin very carefully used to imagine it, and eliminates or transforms the metaphor.

And abstracting from Darwin's language is just what I don't want to do here. The metaphors can't be "skimmed off," for they frequently put the mind back into sentences that are trying to expel it, they tell or imply stories, they give the world an affective dimension, and they produce creative tensions that have something of the instability of the world Darwin teaches us to recognize. Commentators have long recognized the metaphorical nature of his writing, even severely criticizing it in early responses to the *Origin*: such merely literary forms of argument were thought to undercut the scientific authority. In 1866, A. R. Wallace urged Darwin to replace "natural selection" with Spencer's less metaphorical "survival of the fittest," because, as Wallace put it, "this term is the plain expression of the *facts,—Nat. selection* is a metaphorical expression of it—and to a certain degree *indirect & incorrect*, since, even personifying Nature, she does not so much *select* special variations, as *exterminate* the most unfavourable ones." Darwin is yielding and defensive in the face of this suggestion, but hangs on for dear life: "The term Natural Selection," he responds, has now been so largely used abroad & at home that I doubt whether it could be given up, & with all its faults I should be sorry to see the attempt made" (*CCD*, vol. 14, 5 July 1866).

In the fifth edition of the *Origin*, Spencer's phrase does appear, but "natural selection" remains as well, and Darwin's point made to Wallace that it's too bad "survival of the fittest" can't be turned into a noun that governs a verb is just another way for him to say that he needs the metaphor, for he wants a phrase doing the metaphorical work of "selecting" to which Wallace objects, and which he himself will directly disavow. In the Essay of 1844 (published in Francis Darwin (ed.), *The Foundations of the Origen*), Darwin makes plain how critical the idea of "selection" is to his entire argument. It is a metaphor, to be sure, but it does extremely important work, and as Darwin thinks about it, the model of domestic selection *is* at work. In one of his many thought experiments, he suggests: "Let us then suppose that an organism by some chance ... arrives at a modern volcanic island in process of formation and not fully stocked with the appropriate organisms."[19] He then supposes that the new organism would survive even in environmental conditions different from his native ones, but only by changing, by, as it were, becoming "plastic." He explains in some detail that, "as in domestic selection ... every part of the body would tend to vary from the typical form in slight degrees, and in no determinate way, and therefore *without selection* [emphasis in the original] the free crossing of these small variations (together with the tendency to reversion to the original form) would constantly be counteracting the unsettling effect of the extraneous conditions on the reproductive system. Such would be the unimportant result without selection" (p. 87). How to account for the "important" results, that is, speciation, of the kind that Darwin learned had taken place among his finches on the Galapagos? The answer is that the variations are in the end guided by selection: natural selection.

Although in later editions, Darwin finally answered objections to his metaphorizing of the process by providing a literal, abstract but implicitly very angry, "objective" description of what he meant by natural selection,[20] denying that it can consciously "select," the fact that he chose to explain rather than to withdraw it warns us, as Robert J. Richards insists,[21] that much is at stake here. He very much wanted the analogy with "domestic selection," despite Wallace's insistence that domestic selection had nothing to do with what happens in nature. Darwin's metaphorical imagination of the process is part of what produced the idea in the first place and of what made it so effective, and continues to do so.

The experience of reading the *Origin* is profoundly shaped by the dominant metaphor: natural selection chooses for the sake of the good of each organism over which it watches, even while the literal meaning of the metaphor is that all traits not conducive to the survival of the organism (and thus all organisms with those traits) will be killed off. In the early Essay of 1842 (*Foundations*), when Darwin moves from domestic to natural selection, he writes: "Who, seeing how plants vary in garden, what blind foolish man has done in a few years, will deny an all-seeing being in thousands of years could effect (if the Creator chose to do so) either by his own direct foresight or by intermediate means...." (*Foundations* p. 36). The representation of natural selection as a positive and godlike figure is too insistent to be the accidental consequence of the use of metaphor. The metaphor embodies the very way in which Darwin imagined the process. I will, return to this dominant metaphor in more detail in Chapter 3, but here it can suffice to say, in particular if one is ready to read Darwin as a writer rather than as a twenty-first-century evolutionary biologist, that it cannot be "skimmed off." While Darwinian metaphors have evoked rich and sensitive readings in some modern literary analyses, they are normally and quite reasonably set aside by contemporary philosophers and scientists who have other work to do than to regard Darwin as a novelist or poet!

Seeing Darwin perhaps not quite as a novelist but as an imaginative writer who is a true descendant of the Romantic movement in literature and culture, enriches and complicates our understanding of Darwin's "science," as well as our understanding of Darwin's influence on the culture—I will, of course, leave science to the scientists. Although in popular representation Darwin usually looks still to be the jaded, weary, heavily white-bearded sage bringing sad news to the world, wonder is one of the keys to Darwin's prose, as I think it remains, often too secretly, of the work of modern science itself.[22] In this book, as I examine aspects of Darwin's writing and of its influence on later literature, I want to emphasize that wonder. For wonder, as I have claimed, generates everything else, including the most level-headed and rational explanations.

So I am hoping now to build on the work of Beer, Richards, and Holmes in considering Darwin as a writer. Darwin's writing was a condition for his "discovery" of the theory of descent by modification through natural selection but it also evokes another Darwinian reality (content though I am to believe in the reality science is making of his ideas); it is a reality more singular, to be sure, but no less "true" than the reality his theory describes, and no less intensely imagined than the worlds of the great novelists.

Addressing Darwin's writing as literature leads to attention to the many particulars that constitute part of his argument and provide evidence for it; it requires attention also to his modes of expression. The particulars in part resist the large-scale inferences derived from the theory they help to produce; the language that creates Darwin's theory creates also its less generalizable truths, and is infused with Romantic energies that are not all expended in the disenchanting views that are normally attributed to Darwin. Darwin works hard, as I will go on to show in later chapters, to think away the wonder and marvel he feels in contemplating extraordinary or what seem to be naturalistically inexplicable phenomena. But after his remarkable explanations have done their work, readers are left with a sometimes giddying sense of wonder and the complexity and richness and diversity of the strangeness ordinary naturalistic explanations demonstrate.

The danger is that in seeking the Darwin behind the writer who saw nature red in tooth and claw, I will be depicting him (unbelievably) as sentimental and Polyannaish. Anything but. I agree with Cannon Schmitt's judgment that "Recalling the ubiquity of destruction stands as the sine qua non of understanding evolution."[23] I agree with Darwin that a close look at the misery and suffering so intrinsically pervasive in the world and its processes makes it difficult to believe in a benevolent designer.[24] In one of his most famous letters, he wrote to Asa Gray: "I own that I cannot see, as plainly as others do, & as I sh^d wish to do, evidence of design & beneficence on all sides of us. There seems to me too much misery in the world. I cannot persuade myself that a beneficent & omnipotent God would have designedly created the Ichneumonidæ with the express intention of their feeding within the living bodies of caterpillars, or that a cat should play with mice" (*CCD*, vol. 8, 22 May 1860). "There seems to me to be too much misery in the world." Any reading of Darwin's prose that does not come to terms with this aspect of his vision cannot be adequate to his art.[25]

But it is also the a-b-c of natural selection that it builds life out of death—no death, no organic variety, no humans. Although this is in many ways, a very old idea, it has never been an easy one to assimilate, and most certainly not in the context of a Darwinian, fully naturalistic imagination of the world. There must be tension everywhere as Darwin not only loves the world—which he most intensely does—and loves the work of discovering it, but is also repelled by much that goes on in it, from parasites gnawing the insides of their hosts, to savages eating their mothers (well, he got that wrong in the *Journal of Researches*, but that's what he had been told). To achieve his larger argument Darwin knew he had to confront the particularities of living beings of every order, everywhere. In the same letter to Gray, after all, he also wrote, "On the other hand I cannot anyhow be contented to view this wonderful universe & especially the nature of man, & to conclude that everything is the result of brute force." The tension implies high drama. It doesn't *seem* comic in texture; Darwin himself is not happy leaving the working out of natural developments to natural law and chance. But—and here is the critical point that I will be developing throughout this book—finally, *On the Origin of Species* is comic in form. While Schmitt's formulation is certainly correct, so too is the decisive point that follows from Darwin's one long argument: understanding evolution in "this wonderful universe" means also recalling the ubiquity of life. The *Autobiography* registers informally much of the point of my concern here with a "comic" Darwin.

Passing over the endless beautiful adaptations which we everywhere meet with, it may be asked how can the generally beneficent arrangement of the world be accounted for? Some writers indeed are so much impressed with the amount of suffering in the world, that they doubt if we look to all sentient beings, whether there is more of misery or of happiness;—whether the world as a whole is a good or a bad one. According to my judgment happiness decidedly prevails, though this would be very difficult to prove. If the truth of this conclusion be granted, it harmonises well with the effects which we might expect from natural selection. (p. 88)

Note that Darwin does *not* include himself among those who doubt whether the world is a good or bad one. It is certainly a difficult world, full of suffering, but Darwin does not himself "pass over" "the endless beautiful adaptations." The world that he describes, full as it is of suffering and mindless cruelty, is a "good" one, and no doubt he loves its "beautiful adaptations." So the narrative of Darwin's world has comic shape.

But it is comic in another way. Attention to minute, even domestic particulars, is a characteristic of his prose, and a condition for his larger ideas. After all, what makes him disbelieve in intelligent design is a set of small, ordinary things: the Ichneumonidæ, with their hideous mode of feeding within the living bodies of caterpillars, or cats playing with mice. How, in moving toward a world-historical theory can one not find it striking (and even potentially funny) that the movement has to be carried on through attention to things like beeswax, worm castings, and seedlings and—let's call it by its basic name—birdshit?

There are many ways to be Romantic. Two of them are suggested when Carlyle's Herr Professor Teufelsdröckh cries out, "Close thy Byron; open they Goethe"—the former, to emphasize extraordinary events and characters, to go after the big effect, the intense feeling, and to play out one's narrative and even one's sense of oneself against a vast, sublime background; the latter, Wordsworthian (or, in Carlyle's sense, Goethian), to see the world as the English realist novel did, finding feeling imbued deeply in the ordinary—seeing the world in a grain of sand, one might say. Darwin's Romanticism moves between both poles, but pulls always toward the "ordinary," and its transformations. Darwin is Wordsworthian rather than Byronic.[26] Although one might think of Darwin as the adventurer in exotic explorations of the tropics and the most distant parts of the world, even his *Journal of Researches* depends, in a quite Wordsworthian way, on memories of and longings for England, home, and the domestic. The very Antipodes call to Darwin's mind "one's recollection of childish doubt and wonder. Only the other day," he writes during the *Beagle* voyage,

I looked forward to this airy barrier as a definite point in our voyage homewards; but now I find it, and all such resting-places for the imagination, are like shadows, which a man moving onwards cannot catch. A gale of wind lasting for some days, has lately given us full leisure to measure the future stages in our long homeward voyage, and to wish most earnestly for its termination. (p. 416)

Here is an early intimation of a peculiarly Darwinian pattern: the voyage spurred by the lure of the exotic and the tropical now mixes with nostalgia and thoughts of home. The most powerful impetus to his work was the wondrous exotic and adventurous trip around the world, out of which he drew materials and connections to last for a lifetime. Having returned to England, he soon after settled into a home from which he rarely moved for three decades, so that for the rest of his life he was a homebody—or at least, he spent the largest part of his time, when he wasn't too ill, working in his home study, thinking out problems as he strolled the sandwalk, doing little experiments in his home garden, where the world—the botanical world—came to him, playing with his children, being a genial if ailing Victorian paterfamilias. It was from his house and garden—or the mail arriving at his door—that Darwin would draw his materials. The shape of his life and work is strikingly similar to the structure of his arguments and of his theory. His "tropical glow," aroused by Humboldt and poetry even before his voyage on the *Beagle*, is accompanied by a sustained feeling for the ordinary.[27]

Gopnik emphasizes, quite correctly, Darwin's concern with "small things," his emphasis on writing with precision about the most ordinary phenomena: ants; worms; cows; dogs; pigeons; bees; and weeds in his garden. Darwin, Gopnik claims, saw "something large and...found the right words to say it small." He had no pretensions to high literary style and no concern to produce a book that would be beautiful or moving—as, in fact, it sometimes is. He might even have been alarmed if he were to have found himself in this book (as no less in the work of Beer and Hyman) juxtaposed to

Dickens and George Eliot instead of to Cuvier, say, or Lyell. He sought, in an unselfconscious way probably no longer available to modern writers who have learned the craft of negative hermeneutics and have been jaded by suspicion and skepticism after encounters with the trickiness and deviousness of language, to represent things as they are. Darwin, as a writer, was nevertheless self-conscious about this. In a letter to W. H. Bates, he gave some unsolicited advice about writing: "As an old hackneyed author let me give you a bit of advice, viz to strike out every word, which is not quite necessary to connect subjects & which would not interest a stranger. I constantly asked myself, would a stranger care for this? & struck out or left in accordingly.—I think too much pains cannot be taken in making style transparently clear & throwing eloquence to the dogs. I hope that you will not think these few words impertinent. (*CCD*, vol. 9, 25 Sept 1861).

Near the end of his chapter on natural selection, chapter four, after explaining how it works, Darwin writes: "Whether natural selection has really thus acted in nature, in modifying and adapting the various forms of life to their several conditions and stations, must be judged of by the general tenour and balance of evidence given in the following chapters" (p. 127). Characteristically modest and moderate, Darwin here simply asks the reader to attempt to follow his arguments and decide—we provide the facts, you choose! He does provide a cartload of facts, but in asking attention to the "general tenour and balance of evidence" he is appealing to an ideal of what Gopnik calls "liberal" discourse. Does the evidence have sufficient weight? What would a reasonable person, in the face of the evidence that Darwin marshals, have to say? Although he is certainly himself absolutely convinced that natural selection works that way, he understands that his readers need to weigh the evidence themselves, and he knows that the evidence is not *absolutely* clinching. This strategy of liberal openness is not, of course, as disinterested as it seems, since Darwin is aware that if his theory is to be taken seriously, it has to be presented in a way that would persuade the reader. He himself was fully persuaded and pressing as hard as he could to drive the reader to agree. Now, 150 years later, there can be no doubt about the evidence. But Darwin believed in 1858, while knowing that he didn't have the clincher, that he already had enough. Darwin requires of his readers an openness to possibilities, even counter-intuitive possibilities, even possibilities that might make them uncomfortable. It is not only an intellectual experience; emotion, implicit or explicit, will make a big difference. How do you feel about *this*? Look how wonderful *this* is! Registering in his own language the swerves and returns of feeling that any reader might feel in confronting such a nature, Darwin leads us toward another way of looking at the world, a world that positively "vibrates" with meaning and life.

It is a matter as well of recognizing the extraordinary importance and beauty of what Gopnik calls Darwin's "literary eloquence," an eloquence that "is essential to liberal civilization." Darwin matters most, says Gopnik, because he "wrote so well."

"The language" Darwin helped to invent, he claims,

is still a rhetoric that we respond to—a new style of persuasion and argument, that belongs to liberalism. (I mean liberalism here...not in the American sense of well-meaning and wishy-washy, or the French sense of savagely devoted to the free market, but in the British sense, John Stuart Mill's sense, in which an individual is committed, at the same time, to constitutional rule and individual freedom, to the power of the many *and* the free play of the mind—the sense that takes in a "conservative" in our politics as well as a "liberal," if not in a way more.)

Darwin, with the limited equipment available to him, made the case extraordinarily well in a prose that we can learn from and feel. My only small quarrel with Gopnik's way of valuing Darwin's writing is rather an amendment: along with this "liberal" and reasonable discourse, Darwin's writing almost always carried that affective freight that I've already intimated. It is rather more imaginative and speculative than this Enlightenment version of his practice suggests, and in later chapters I will emphasize the way

the imaginative aspects of the work did have important consequences for other writers. His arguments have the power that
Gopnik so well describes; but they have another, what I would
like to call "writerly" power, that moves ideas into experience and
weights them with the strength of feeling and imagination, not
least in the use of metaphors, but equally in what looks like careful
recording of the facts.

His is a rhetoric designed not to displace rational discourse, not
to deceive, but somehow, out of the depths of his own personal
feeling and talents, to find its way toward an understanding of how
the world really works. As Adam Phillips has put it, Darwin
thought of himself "as trying to tell the truth about nature, and
nature was what truth was about."[28] His prose then is marked by
a dogged openness to fact, a kind of loving submission to detail, an
insistence on representing—even at the expense of tedium or, as
he called them, "dry facts" as accurately and honestly as possible
the phenomena he describes, and it is to make no distinction
between the great and the small. But it also bears the marks of a
passion for the nature it describes. Darwin shows himself impa-
tient with the most direct and abstract uses of thought, the kinds
that make the rapid inference that, for the most famous example,
design implies a designer. His passion for the truth and openness
leads him beyond theory into the intricacies of individual organisms, so that the reasoning really can begin only after saturation in
the particulars, which have their own remarkable affective power.
And yet, and at the same time, Darwin's "submission" to facts is
consistent with his constant reorganization of them until they
produce a convincing narrative. Facts are never single, always in
context, and often Darwin has to imagine a context that is not
immediately visible, though it is one, he would try to convince his
readers, that worked inside our normal sense of cause and relationship. And thus, for Darwin, the representing of nature accurately meant also the representation of what was not immediately
there, and required great leaps of imagination.

One representative but particularly stunning example of this comes in his elaborate discussion of ants in his chapter on "instincts." On the face of it, instinct poses the greatest threat to his theory, and the threat is intensified by the existence of sterile worker ants, which cannot, by definition, pass their characteristics on to future generations. And so Darwin looks intensely at the differences between and within different castes of ants—the "Eciton," among whose "soldier neuters" there are many with "jaws and instincts extraordinarily different," or the "Cryptocerus," in which "the workers of one caste carry a wonderful sort of shield on their heads," or then, the Mexican Myrmecocystus, whose workers "never leave the nest," fed by other workers who have "an enormously developed abdomen which secretes a sort of honey," and so on into intricate relations and further ranges of difference.[29] If we, as lay readers, can maneuver through the difficult names of these ants—names that, once registered, might provide their own sort of pleasure to the reader—we may follow Darwin through these details into the arguments that make all this count for evidence for his theory. It was critical to his theory that he explain how these insects that did not reproduce themselves could carry their characteristics into the next generation, and so he had to make a leap beyond mere description, beyond the precise record of their behaviors, to large inferences that will make these behaviors consistent with the overall theory of natural selection.

He does indeed make that leap and explains how characteristics can be transmitted without direct procreation by the creatures that have them. But while he does that triumphantly, the fact of the extraordinary particularity and diversity, and the fascination of the minutiae, and, for example, that "wonderful" inexplicable shield on some Eciton workers, survives the argument in its very particularity.[30] Follow Darwin into the details of the nature he observes, in its minutest particulars, and wonders will emerge. Darwin, as Phillips says, in telling the truth about nature, tells the truth about its details, not only about its inferable processes. We can understand the explanation and be moved by it as well.

But beyond the intensity of detail there is another characteristic of *On the Origin of Species*, to which I have already alluded and which, in a way, should be present in the background of everything that might be said about the effectiveness (and affective-ness) of his prose. It is one that almost invariably surprises new readers: the quality of "personal testimony," a voice, however often muted, of a personality, and the intimation in it—even if the intimation is not officially part of the overall argument—of a narrator who loves the natural world and who finds it not only in exotic places, but at home. We can hear just a gentle reverberation from it in the passage about ants. The *Origin*, though aspiring to universality and objectivity, is also a surprisingly personal and domestic book, in which Darwin emerges as a man aware of the difficulties of his position but ultimately confident that it is right. For example, after showing the readers the extraordinary diversity of worker ants, he emerges in his own voice: "It will indeed be thought that I have an overweening confidence in the principle of natural selection, when I do not admit that such wonderful and well established facts at once annihilate my theory" (p. 239). So he enacts the readers' doubts, makes clear his awareness of the difficulty, parades an awkward discomfort about his "overweening" confidence, and then proceeds in great detail to show why he does indeed have that confidence. This is his strategy here and throughout the *Origin*. The difficulty once registered, he intimates a self who almost embarrassedly realizes how cocky he must seem, and then proceeds to persuade by plunging into the minutest particulars reflecting the minutest attention and speculating daringly about them. The passages themselves become tours de force, almost magical explorations as well as examples of intricate reasoning. The several pages he devotes to ants conclude with Darwin prov-ing that what seemed to be a "difficulty" for his theory in fact turns out to be its richest confirmation. He shows that those amazing differences among castes are extremely useful (there is a metaphor here of the modern production line that in itself deserves atten-tion). And he concludes, dazzlingly, making the ants' sterility essential for their success: "a perfect division of labour could be effected with them only by the workers being sterile; for had they been fertile, they would have intercrossed, and their instincts and structure would have become blended. And nature has, as I believe, effected this admirable division of labour in the communities of ants by means of natural selection" (p. 242). The rhetoric is virtually sublime in its quietly developed confidence emerging through attention to minutiae and sweeping away the difficulties that had initially threatened to "annihilate" his theory. Darwin finds a way to persuade with a kind of bold honesty that makes his own original doubts part of the argument: "I should never have anticipated that natural selection could have been efficient in so high a degree, had not the case of these neuter insects convinced me of the fact" (p. 242).

Darwin's humility and honesty work brilliantly here almost as a kind of Wordsworthian egoistic sublime, moving from the ordinary against his own original instinct to achieve extraordinary effects. And throughout the *Origin*, his garden and its weeds and flowers figure almost more prominently than the exotic lands of so many of his examples. He does "little experiments," gets "three tablespoons full of mud" from different spots on a local pond, weighs the mud, keeps it for six months, and counts the weeds that emerge—537, he claims, with an exclamation point, all in a breakfast cup! (p. 387). The spoon and the breakfast cup do not make the implications of the very large number any less forceful scientifically, but they remind us, even if that's not Darwin's point, of a domestic Darwin, and prepare us for an understanding that the domestic and the world-historical argument are tightly implicated. What has a breakfast cup to do with the question of whether seeds can survive long periods and still grow, or whether they can "move," and find some means to be transported from one area to another? Breakfast cup *and* tropical jungle: they are both objects of wonder. Darwin asks us to think through juxtaposition—if we get that many weeds from three spoonfuls of mud in the cup, what might birds, whose feet are often muddy, do in the movement of plants across large distances? Spoon, cup, muddy-footed birds—big theory! At the beginning and the end, as things get explained in the most reasonable terms, there is wonder and astonishment—so much life in a breakfast cup, so astonishing a process to be inferred from it.

The experience of evolution as it emerges from the *Origin* is therefore not some remote abstract and theoretical one. Darwin's prose brings it home to our bodies and our gardens and our zoos and wherever in our ordinary lives we encounter forms of life. One hears often behind the prose, which is sometimes, it is necessary to admit, a bit clumsy in its passion to get it precisely and objectively right, that Darwinian personality: modest, hard-working, ready to get his hands dirty, planting cabbages or digging weeds, warmly committed to his subject in a way that it is hard not to feel as endearing. And one who, though obviously very conscious of what it takes to build a strong argument, seems rather unselfconscious about the intensity of his engagement with the world and about the way his thinking and observing transform his rendering of it into something rather more resonant and full-bodied than one might expect from the clearly argued prose for which he worked so hard.

This Darwin is not the one who normally gets into the papers, or who has been much detected in literature ostensibly influenced by him. The normal reading of Darwin and his influence was formulated very well, long before Gillian Beer's groundbreaking study of Darwin in 1983, in Hyman's still important, *The Tangled Bank*, which treats Darwin as a writer (just as I am proposing to do here) and which notes how remarkable it is "to find that *The Origin of Species* is a work of literature" (p. 34). But *The Origin of Species*, claims Hyman, has "the structure of tragic drama" (p. 34), a plausible and well-argued claim. Obviously, many writers, not least Thomas Hardy, to whom I will be turning in later chapters, felt the tragedy in Darwin's way of representing the world. But he felt much else, as well. After Beer's more complex analysis of Darwin's writing, it has become equally possible to think of alternative Darwins. It is possible to make the claim to which much of this book will be devoted, and which is borne out not only by Darwin's prose but by the work of many who were profoundly affected by what he wrote: the *Origin* has the structure of comedy. It works through the multitudinous details of the "struggle for existence" to emerge, as the last paragraph famously does, claiming that "there is grandeur in this view of life."

I don't want to underplay the tragedy, or the universality of destruction that Schmitt invokes, but I want to suggest here why the prose of the Wordsworthian Darwin makes the world "vibrate" for Gopkin, why it fills the world, both for Darwin and for me, with wonder, and why it refuses to settle into the meaningless bleakness and the mindless disaster that so many writers following Darwin at least partially learned from him. Adam Phillips has also rejected the idea that Darwin was a pessimist, despite the centrality to Darwin's vision of death and destruction. As he puts it, Darwin is only a pessimist "compared to certain previous forms of optimism (the belief in redemption, or progress, or the perfectibility of Man). We are not merely trapped in . . . nature, we are also released into it." We read him for his "redescription of happiness" (p. 12). The world is brutally hard, yes; it is unfair, yes; extinction is not the exception but the norm, and is, given enough time, inevitable, or almost. Darwin himself once wrote: "What a book a devil's chaplain might write on the clumsy, wasteful, blundering, low, and horridly cruel works of nature."[31]

Yet Darwin struggled, if not always, as Monte Python would have it, to "look on the bright side of life." At the end of his chapter on "The Struggle for Existence," he is working very hard at it. Perhaps too hard, since the concluding sentence is relatively feeble and somewhat unconvincing (one of Darwin's lesser achievements as a writer, I would say, and yet authentic), perhaps a sort of whistling in the dark. It is significant that he does pause to face directly what he also must be feeling to be the threatening darkness of the world he has been describing. In any case, he claims, clearly aware of how dark the vision might be taken to be, that "When we reflect on this struggle, we may console ourselves with the full belief, that the war of nature is not incessant, that no fear is felt, that death is generally prompt, and that the vigorous, the healthy, and the happy survive and multiply" (p. 79) There is something clean and economic about this sentence, significantly void of the "wonder" that marks so much else in the *Origin*. But Darwin was, as I say, working at it.

As many have noted, this way of thinking did not console him at the death of his beloved daughter Annie, when she was only ten. But while most commentators emphasize Annie's death, few build on the crucial fact that another child, Horace, was born three weeks afterwards. Life out of death. The argument is barely convincing. The reality seems to have been that even for Emma, "a new baby can never obliterate the gap left by a dead child."[32] Nevertheless, this effort and redemption of death through life sets a *pattern* for Darwin's argument throughout his work, and it infuses itself, in different ways, into the very texture of his prose. At the level of the detail, this pattern usually successfully implies a tension between alternative understandings and thus another way to feel about the world. Confronted time and again by an overwhelming and adamant fact whose interpretation seems obvious and to which only an expression of awe seems adequate, Darwin finds ways around it—explanations that shift our sense of the fact. We have seen that pattern worked out in his discussion of worker ants. Closing his Byron, he opens his Wordsworth. If there is struggle everywhere always, there are ways to understand that struggle that alter its significance. Life out of death. Growth out of struggle. If the passage at the end of "The Struggle for Existence" has something less than the fully consolatory about it, if the very fact that "consolation" seems necessary is a bit depressing, it is nevertheless the mark of a struggle to find language adequate to a full expression of a vision that will not go as dark as it is popularly taken to be.

Darwin is looking for the other side of the "tragic vision" that Hyman describes. Darwin had to negotiate his way through extinction and the pervasive indifference of nature, which he did not try to minimize (and were in fact essential to his theory), but without losing the sense of the wonder, beauty, and excitement of life in all its variations, as it followed from death. The vision is paradoxical, and it is the figure of paradox on which I will be focusing in my efforts to understand how Darwin's prose works. The paradoxical vision marks the larger arguments of the *Origin*, which is, in one of its aspects, an enormous compilation of details about the varieties of life on this planet. The reversal implicit in "life out of death," in the mindlessness of what seems mindful, in the counter-intuitive perceptions and inferences that emerge everywhere, in virtually every discovery and manifestation of life in Darwin's work, leads to what I will be calling the "double movement" of his prose, a movement that allows him to register rather unscientific, intuitive feeling, to struggle with that feeling, and to emerge again, this time with new scientific knowledge, but with the feeling altered yet intact. In Chapter 3, I will look more closely at this pattern and attempt to trace in detail through recurring passages in the *Origin*, the characteristics of his prose that I have been discussing here.

But the something larger that Darwin digs out of the small, the truth about nature, turns out to give weight to Beer's point that his theory was laden with "the remnant of the mythical" (*Darwin's Plots*, 2). The truth takes the shape of a story, and it is the story of our world as a kind of secular myth—though, as Beer also points out, "Evolutionism has been so imaginatively powerful precisely because all its indications do not point one way" (*Darwin's Plots*, 6). The *Origin* is a prosy *Divine Comedy* or "Paradise Lost," an alternative myth to the story that begins with creation and "man's first disobedience," and it implies far more than Darwin himself could have known. To rank Darwin among the great writers is no tricky trope, but a fact of his labors: he *was* a writer, one who, as Beer puts it, was "telling a new story, against the grain of the language available to tell it in" (*Darwin's Plots*, 3). In literary study, students follow Milton's self-conscious development into an epic poet and into a choice of topic that would have the widest epic significance—and he couldn't have chosen a larger one. It is possible to watch just such a development in Darwin, who, despite his modesty, also had Miltonic ambitions in choosing to pursue what John Herschel called the "mystery of mysteries." The Bible tells its story; Dante, Milton, Darwin tell theirs.

In locating Darwin's writerly qualities, and the particular nature of his writing that makes reduction to a "tragic vision" inadequate, I am following not only Beer, Richards, and Gopnik, but an excellent if insufficiently attended to essay, written many years ago, by A. Dwight Culler, who, it seems to me, captured an essential quality of Darwin normally unnoticed by critics who cannot resist seeing nature red in tooth and claw, and locating Darwin where things are most gloomy, and most serious. The form of Darwin's argument, Culler claimed, is more important for our understanding of the relation of his work to the art that followed it than the apparent thematic content. "To trace the way in which ... writers derive from Darwin their views of nature, man, God and society," wrote Culler, "does not seem to me quite to get at the heart of the problem."[33]

While Hyman had insisted that the form of the *Origin* is "tragic," Culler saw something in Darwin that entirely changes the terms most appropriate for talking about his writing and his influence on other writers. It is a given that Darwin was intent on reversing the arguments of natural theology. Nothing is more rational than the idea that an apparently designed thing must have a designer; nothing is further from what Darwin tells his readers he saw in nature. And thus, everything in Darwin turns out to be counter-intuitive: what *seems*, to which Darwin in his role as natural historian attends with the greatest care, always disguises what *is*. And, Culler claims, the form of argument that most clearly can express this view is the form of paradox and irony. Adaptation is not the result of the work of a guiding Intelligence but of chance (although, to make things even harder, Darwin did not believe in "chance"). So the tragic vision becomes, rather, the ironic or paradoxical one. The world, in one sense, becomes a great joke on the purely rational.

Where most taxonomists studied species by focusing on the similarities, Darwin found that the exception, the anomaly, was most telling and important. "Darwin," claims Paul Barrett, "was always looking out for natural phenomena that would be imperfect or pointless from the point of view of an all-knowing Designer."[34] Just as Victorian realism attempted to elevate the ordinary and the singular to primary importance and give to that ordinariness the value traditionally associated with the extraordinary, so Darwin depends on the singular and aberrant to guide us back to larger meanings. What doesn't seem important, leads Darwin more directly to the meaning of things than what does. The Darwinian aesthetic, then, leans toward parody and paradox, repeating the traditional forms and reversing the meaning. It looks at what doesn't seem to fit, at what doesn't belong, foregrounds it, makes us aware of it, and in so doing transforms everything.

Darwin's world is notoriously a mixed world, one in which categories normally viewed as distinct are intimately connected. And the re-reading of Darwin suggested by Culler is amplified by Jonathan Smith, who writes,

Darwinism, with its blurring of boundaries and blending categories, its focus on variation, eccentricity, and irregularity, and its interest in bodies and their functions, especially sexual reproduction, digestion, defecation, and death,... had many of the hallmarks of the grotesque. While Darwin and his allies clearly reveled in nature's grotesqueness, and took delight in their celebration of them, those who regarded nature as complete, ordered, stable, and hierarchical looked upon Darwin's vision with horror.[35]

It is not only that the Darwinian method leads to inversions and paradox, but that it points to a generically mixed mode, in a world where values are turned upside down (since Darwin was to argue for the material basis of both art and morality) and where conventional proprieties about the body, and expectations of unified identity, consistent behavior and feeling, are constantly disappointed. The grotesque and the sometimes horrifying recur in Darwin, but the form, the upsetting of expectations, the juxtaposition of unlikely objects, the turn from the universal to the singular (and back again)—all of these moves have the *form* of a joke, where the rational and the expected is displaced.

Much of Darwin's work, claims Smith, "can be characterized as grotesque." And he lists, along with the fascination with worm castings and their work, "the bizarre sexual arrangements of barnacles and orchids; the outré forms of fancy pigeons; the extravagant plumage, ornament, and weaponry of male birds; the hideous facial expression [in illustrations for *The Expression of Emotions*] of Duchenne's galvanized old man; the elaborate traps of insectivorous plants."

The grotesque was indeed, as Gillian Beer pointed out, a major motif of Victorian culture (p. 81)[36] and the Victorian penchant for the grotesque, detectable in Ruskin, and Dickens and Browning, among the most famous of its cultural stars, and in the elaborate furniture and decorations that one thinks of as characteristically Victorian, was everywhere. And yet the fascination of the grotesque, until at least its fullest blossoming at the *fin de siècle*, tended to be largely compatible with the Victorian rage for order against the blooming and threatening new multiplicities of culture, empire, and class. We can watch it in Carlyle's often grotesque prose as it aspires to a "pole star" that can guide us comfortably through the morass, and we can see it in Ruskin's hostility to the grotesqueries that were part of Darwin's vision, even while Ruskin himself celebrated the grotesques of gothic architecture. Darwin's exploitation of the grotesque was a function of the nature of his task and his argument: the grotesque, it developed, was the best argument against the idea of a world entirely and rationally designed, and nature was full of grotesques.

I would add to Culler and Smith's arguments, the other aspect of Darwinian prose that implies the double nature of vision, the juxtaposition of the vast with the domestic, of the rational with the irrational, of the ordered, and the grotesque. We have seen some of that in Darwin's discussion of ants. When Darwin begins to engage a subject in nature, his response is just that expression of Romantic awe that I've been calling "wonder." Or he finds himself intimidated and puzzled by some natural phenomena that seems, on first view, to belie his theory—how can natural selection possibly account for this awesome thing? He feels it as he confronts the eye of the eagle, the comb of the honey bee, the slave-making and sterile ants, the neuter drones of bees, the intelligence of the human, and yet he looks on with "enthusiastic admiration" (p. 224), at these "truly wonderful facts" (p. 128). He begins his engagement with "difficulties on theory," for example, "by conceding that these difficulties "are so grave that to this day I can never reflect on them without being staggered." So Darwin begins "staggered," as elsewhere he begins in awe, apparently doubtful that explanation of these extraordinary phenomena are compatible with scientific reason. But, he says, in a characteristic second move, "to the best of my judgment, the greater number [of difficulties] are only apparent, and those that are real are not, I think, fatal to my theory" (p. 171). What follows, then, time after time, is a patient, detailed slow working out in which what began as "staggering" ends by being naturalistically explicable in terms of ordinary causes now in operation, and not only explicable but, as in the case of the ants, absolutely essential. We will watch this pattern emerge at almost every crux in Darwin's arguments. This is the "double movement" of his prose.

The final chapter of the *Origin* begins working the same pattern: "Nothing at first can appear more difficult to believe than that the more complex organs and instincts should have been perfect, not by means superior to, though analogous with, human reason, but by the accumulation of innumerable slight variations, each good for the possessor" (p. 459). Yet, of course, Darwin will overcome our "imagination," to which this difficulty seems "insuperably great." He moves through a set of propositions that "cannot, I think, be disputed." Recapitulating, he deals with objection after objection, impossibility after impossibility, working on our "imagination," and arrives at what happens "under domestication" (p. 456). Brought home again to the ordinary, the knowable, Darwin makes the connection once more: "There is no obvious reason why the principles which have acted so efficiently under domestication should not have acted under nature." The wonderful, the incredibly difficult, the insuperable, become ordinary. And all the more wonderful.

Notes

1. For a detailed history of the publication record of *Vestiges* see James Secord, *Victorian Sensation: The Extraordinary Publication, Reception, and Secret Authorship of* "Vestiges of the Natural History of Creation" (Chicago: University of Chicago Press, 2000), esp. ch. 4. Secord's remarkable study makes plain that Chambers was more important than Darwin in softening up a culture for the idea of evolution; his book took the brunt of the attack, in particular from scientists themselves. There is no doubt that the idea of evolution was more readily accepted by a culture that had found Chambers's book so fascinating and attractive, but Chambers' imagination of the evolutionary process is distinctly not that of Darwin.

2. Gillian Beer, *Darwin's Plots: Evolutionary Narrative in Darwin, George Eliot, and Nineteenth-Century Fiction*, 3rd edn. (Cambridge: Cambridge University Press, [1983] 2009).

3. Stanley Edgar Hyman, *The Tangled Bank* (New York: Grossett and Dunlap, [1959]1966), 35.

4. In thinking through what I will be calling a lack of precise fit between Darwin's writing and the theories they so powerfully represent, I have been greatly helped by the work, and talk, of Myra Jehlen, who writes about literature and art as modes of knowledge, if precisely *not* the kind of knowledge that can be developed into generalized and regular laws. Her subtle and fine analyses are best represented in her recent, *Five Fictions in Search of Truth* (Princeton: Princeton University Press, 2008). Averse to explicit theorizing, Jehlen comes closest to explaining this conjunction of the epistemological and the aesthetic, discussing, but not overtly theorizing, the way particular works of art produce particular knowledge *not* reducible to the general, but nonetheless "objective," and experientially real. See "On How, to Become Knowledge, Cognition Needs Beauty," *Raritan* 39: September, 2010, 39–46.

5. *The Correspondence of Charles Darwin* (henceforward *CCD*), ed. Frederick Burckhardt et al. vol. 9 (Cambridge: Cambridge University Press, 1994), to H. W. Bates, 3 Dec 1861, 363.

6. Adrian Desmond insists that we should not "believe those who tell you that [the *Origin*] is brilliantly written." Adrian Desmond, *New York Times Book Review*, 27 August, 2006.

7. For any reader interested in getting clear exactly what the *Origin* said and didn't say, how Darwin got to his ideas, what the difficulties and strengths of his argument were, there is a wonderful collection of essays on the relevant subjects in *The Cambridge Companion to the* "Origin of Species," eds. Michael Ruse and Robert J. Richards (Cambridge: Cambridge University Press, 2009). Many of them get far more technical than this book, but all are written in a way accessible to the non-scientific reader, and one, by Gillian Beer, does address the subject from the perspective of literary study.

8. After most of this book was written, I was introduced to a splendid 1946 essay by Theodore Baird, "Darwin and the Tangled Bank," originally published in *The American Scholar* and reprinted in a collection of Baird's essays, *The Most of It*. (I have only seen a Xerox of the essay and thus cannot provide bibliographical information on the book, which doesn't appear on the normal lists.) Baird not only makes the case for Darwin's work as "literature," but (not quite to my dismay) he singles out passages from Darwin's *Journal of Researches* on which I had focused in the second chapter of this volume, and towards the end, makes the point that "the emotions expressed so modestly in *The Voyage of the Beagle* recur in more generalized form in the *Origin*. Wonder and amazement predominate" (p. 171).

9. Beer's *Darwin's Plots* set the standard for treatment of Darwin's metaphors. She locates in Darwin's work an "impulse to substantiate metaphor and particularly to find a real place in the natural order for older mythological expressions" (p. 74). Hyman talks of Darwin as a "metaphorist" organizing all his most important ideas as metaphors around the central one of "natural selection" (p. 34). See also David Kohn, "The Aesthetic Construction of Darwin's Theory," *Aesthetics and Science: the Elusive Synthesis*, ed. A. Tauber (Dordrecht: Kluwer, 1996), 13–48. Discussion of Darwin's metaphor, even among those not directly concerned with his "writing," are pervasive in the Darwin bibliography.

10. David J. Depew, *The Rhetoric of the* Origin of Species, in Michael Ruse and Robert J. Richards (eds.) *The Cambridge Companion to the* "Origin of Species" (Cambridge: Cambridge University Press, 2009), 238. In this excellent essay, Depew connects some of the strategies of Darwin's writing, which I will

be discussing in different terms, to traditions of rhetoric about which Darwin had read.

11.   Inspired by reading von Humboldt, Darwin was planning to go to the Canary Islands in June, 1832 before he got the wonderfully timed invitation to join the *Beagle* voyage. His letters in 1831 are sprinkled with allusions to von Humboldt and the planned trip. To his sister Caroline, he wrote on 28 April, "my enthusiasm is so great that I cannot hardly sit still in my chair ... I have written myself into a Tropical glow" (*CCD*, vol. 1, 122).

12.   David Kohn, "The Aesthetic Construction of Darwin's Theory," in A. I. Tauber (ed.), *The Elusive Synthesis: Aesthetics and Science* (Netherlands: Kluwer Ac. Pub., 1996), 13–48.

13.   No consideration of the nature of Darwin's writing can omit consideration of Gillian Beer's groundbreaking, *Darwin's Plots: Evolutionary Narrative in Darwin, George Eliot and Nineteenth-Century Fiction*, now in its third edition (Cambridge: Cambridge University Press, 2009. The most impressive case for Darwin as a Romantic is made in Robert J. Richard's, *The Romantic Conception of Life* (Chicago: University of Chicago Press, 2002).

14.   Richard Holmes, *The Age of Wonder: How the Romantic Generation Discovered the Beauty and Terror of Science* (New York: Pantheon Books, 2009).

15.   *The Voyage of the Beagle*, ed. Leonard Engel (New York: The Natural History Library, Anchor Books, [1839] 1962), 500.

16.   *The Autobiography of Charles Darwin, 1809–1882* (New York: W. W. Norton, 1969, 1958), 91.

17.   Charles Darwin, *The Formation of Vegetable Mould through the Action of Worms with Observations on Their Habitats* (*London: John Murray*, 1883), 308.

18.   *Romantic Conception,* 537. Richards also devotes much of his essay in the *Cambridge Companion* to this subject. See "Darwin's Theory of Natural Selection and Its Moral Purpose." "Darwin recognized, if dimly, that his original formulation of the device and cognitively laden language of his writing carried certain consequences with which he did not wish to dispense—and, indeed, could not dispense with without altering his deeper conception of the character and goal of evolution" (p. 64).

19.   *The Foundations of The Origin of Species*, ed. Francis Darwin (Bibliobazar, [1909] 2008), 86.

20.   In later editions, Darwin tries to fight off some of the misreadings of natural selection resulting from its metaphorical work: "Several writers have misapprehended or objected to the term Natural Selection. Some have even

imagined that natural selection induces variability, whereas it implies only the preservation of such variations as arise and are beneficial to the being under its conditions of life. No one objects to agriculturalists speaking of the potent effects of man's selection, and in this case the individual differences given by nature, which man for some object selects, must of necessity first occur. Others have objected that the term selection implies conscious choice in the animals which become modified; and it has even been urged that, as plants have no volition, natural selection is not applicable to them! In the literal sense of the word, no doubt, natural selection is a false term; but who ever objected to chemists speaking of the elective affinities of the various elements?—and yet an acid cannot strictly be said to elect the base with which it in preference combines. It has been said that I speak of natural selection as an active power or Deity; but who objects to any author speaking of the attraction of gravity as ruling the movements of the planets? Every one knows what is meant and implied by such metaphorical expressions; and they are almost necessary for brevity. So again it is difficult to avoid personifying the word Nature; but I mean by Nature, only the aggregate action and product of many natural laws, and by laws the sequence of events as ascertained by us. With a little familiarity such superficial objections will be forgotten."

21. Richards wants to claim that Darwin's entire conception of "natural selection" included a sense of moral direction that has entirely dropped out of modern understanding, which picks up the notion that Darwin often quoted from de Candolle, that nature was at war. Richard insists that Darwin's early belief in progress, perfection, and Divine direction lingered vestigially in the idea of natural selection (to which, of course, I will return in later chapters), which is, as Richards puts it, a "divine surrogate." See, "Darwin's Theory of Natural Selection and its Moral Purpose," in *The Cambridge Companion to the* "Origin of Species," 64.

22. Even the anti-sentimental, hard-headed, and rhetorically uncompromising Richard Dawkins devotes an entire book to that experience of "wonder," invoking a remarkable range of poetry along the way in his *Unweaving the Rainbow: Science, Delusion and the Appetite for Wonder* (New York: Houghton Mifflin, 1998), a book, Dawkins claims, that was "inspired by a poetic sense of wonder" (p. xii).

23. Cannon Schmitt, *Darwin and the Memory of the Human: Evolution, Savages, and South America* (Cambridge: Cambridge University Press, 2009), 157.

24. In this volume I will not be dealing with the larger question of the compatibility of Darwin's thought with religion or whether he remained religious himself. It is sufficient to note that he always sought, in his science,

for explanations confined to the natural world. It seems likely that his ultimate description of himself as "agnostic" was accurate, but that he retained a religious feeling (that was, however, severed from any Christian dogma and the Christian narrative of creation and resurrection). For a fine overview of the question, see John Hedley Brooke, "'Laws impressed on matter by the Creator'? The *Origin* and the Question of Religion," in *The Cambridge Companion to the* "Origin of Species" eds. Robert J. Richards and Michael Ruse (Cambridge: Cambridge University Press, 2009), 256–74.

25.   In a touching and quite open sequence in his *Autobiography*, in which he discusses the attributes of an omnipotent God (omitted from the original publication by his wife, Emma, who remained very pious all her life), Darwin writes: "it revolts our understanding to suppose that his benevolence is not unbounded, for what advantage can there be in the sufferings of millions of the lower animals throughout endless time?" (p. 90).

26.   In what can now be seen as a prescient essay, well anticipating the work of Gillian Beer, Marilyn Gaull locates Darwin's romantic way of seeing and writing in relation both to contemporary geology and to Wordsworth's poetry, which we know Darwin read, and not only *after* his return from the *Beagle*, when he read The Excursion twice. See "From Wordsworth to Darwin: 'On to the Fields of Praise,'" *The Wordsworth Circle*, 10(1) (Winter, 1979, vol 10, no 1).

27.   For my argument, made in different ways by Gillian Beer, that Darwin's writing was harmonious with the tendencies of Victorian realist fiction, see my *Darwin and the Novelists* (Cambridge, MA: Harvard University Press, 1988).

28.   Adam Phillips, *Darwin's Worms* (New York Basic Books, 2000), 11.

29.   *On the Origin of Species* (Cambridge, MA: Harvard University Press, 1964), 238.

30.   For a detailed and exhaustive discussion of the explanation, see Robert J. Richards, *Darwin and the Emergence of Evolutionary Theories of Mind and Behavior* (Chicago: University of Chicago Press, 1987). Richards explains simply that Darwin came to recognize that the "principle of selection" is "not of the individual which cannot breed, but of the family which produced such individual." These are Darwin's words from his big "Species Book." Here is Richards' interpretation: "if a community of ants...happened to produce neuters whose structure and instincts benefited the group as a whole, the nest would have a competitive advantage of the other nests and would hence be selected" (p. 150). Reading the complex history of Darwin's

thought on this issue in Richards' recounting is yet another reminder of the remarkable power of Darwin's prose to manipulate complexities and make them available, even to a lay reader.

31.   Darwin to J. D. Hooker, 13 July, 1856. *The Correspondence of Charles Darwin*, vol 6, 1856–57, eds. Frederick Burckhardt, et. al. (Cambridge: Cambridge University Press, 1990), 178. Richard Dawkins, who always emphasizes the mindlessness and brutality of nature, entitles one of his books, a collection of some of his fugitive essays, *A Devil's Chaplain*, but interestingly, seeming to push the theme of the alternative reading of Darwin that I want to emphasize here, he subtitles it, *Reflections on Hope, Science, and Love* (New York: Houghton Mifflin, 2003). Dawkins relentlessly insists that the joy of science comes only through and after the acquisition of real knowledge, and indulges an heroic rhetoric. Here is his version: "Safety and happiness would mean being satisfied with easy answers and cheap comforts, living a warm, comfortable lie. The daemonic alternative urged by my matured devil's Chaplain is risky. You stand to lose comforting delusions: you can no longer suck at the pacifier of faith in immortality. To set against that risk, you stand to gain 'growth and happiness'; the joy of knowing that you have grown up, faced up to what existence means, to the fact that it is temporary and all the more precious for it" (p. 13). I fear Dawkins would consider the Darwin I am trying to invoke in this book through a reading of his prose rather closer to a comfortable lie, though perhaps not yet evidence of thumb-sucking.

32.   Janet Browne, *Charles Darwin, a Biography: Voyaging*, volume 1 (New York: Knopf, 1995), 505.

33.   "The Darwinian Revolution and Literary Form" in George Levine and William Madden (eds.), *The Art of Victorian Prose* (New York: Oxford University Press, 1968), 225.

34.   Charles Darwin, *Metaphysics, Materialism, and the Evolution of Mind*, trans. and anno. Paul H. Barrett (Chicago: University of Chicago Press, 1974), 66.

35.   Jonathan Smith, *Seeing Things: Charles Darwin, John Ruskin, and Victorian Visual Culture* (Cambridge: Cambridge University Press, 2007).

36.   See Beer, p. 81. For an excellent collection of essays theorizing and discussing a great variety of Victorian grotesque writing and art, see David Amigoni, Paul Barlow, and Colin Trodd (eds.), *Victorian Culture and the Idea of the Grotesque* (Aldershot: Ashgate, 1999). In an introduction, the editors point out many of the contradictions that the grotesque implied, in its great popularity, in its emergence in science (see Smith, cited above), and in a wide range of Victorian artifacts, a counter-thrust to dominant Victorian commitments

to order and stability. The grotesque was disrupting and fascinating for the Victorians. In an essay about Darwin and other writers and artists, Nicola Brown points out how Darwin asks his readers to admire the beauty of the world, but a beauty that is not in "harmony, proportion, agreement, symmetry or design," but rather in "chance, and change, mutation and struggle" (p. 120).

2

Learning to See: Darwin's Prophetic Apprenticeship on the *Beagle* Voyage

I shall always feel respect for every one who has written a book, let it be what it may, for I had no idea of the trouble, which trying to write common English could cost one.

CCD 7 July 1837 v. 2, p. 29

If I live to be eighty years old I shall not cease to marvel at finding myself an author.

v. 2, p. 53

I

Although for Darwin, writing was not easy, he had every reason to marvel, for he *was* an author. He often found the work tedious, and clearly he preferred being out in nature and studying it than inside—whether in the sooty city or his almost pastoral home retreat at Down—scribbling away about it. Nevertheless, he almost never stopped writing, beginning his journal (which, as he told his sister "is not a record of facts but of my thoughts") at the outset of his voyage on the *Beagle*. Along the way,

he compiled something like eighteen other notebooks on special-
ized subjects, and, on his return, plunged into what was to be the
third volume in a series about the *Beagle* voyage. This he described
to his friend, William Fox:

a kind of journal of a naturalist, not following however always the order of
time, but rather the order of position.—The habits of animals will occupy a
large portion, sketches of the geology, the appearance of the country, and
personal details will make the hodge-podge complete" (*CCD* vol 1, 12 March
1837, p. 11).

The hodge podge, which turned out to be the *Journal of Researches
into the Geology and Natural History of the Countries visited
during the Voyage of H. M.S. Beagle round the World, under the
Command of Capt. FitzRoy, R. N.*, became an immediate success,
leading to a second edition (the one most widely available these
days), detached from the official three volumes recording the
Beagle's adventures and discoveries.[1]

    With all his natural modesty, there can be no question that
Darwin was always writing something. He was an obsessive note
taker, describer, letter writer, and diarist. Certainly, he took pride
in his writing, and he claimed in his *Autobiography* that "The
success of my first literary child always tickles my vanity more
than that of my other works" (p. 116). There are reasons that it
became a popular book, not least because it was partly written out
of the "tropical glow" into which von Humboldt had led him, in
part because it was written in a genre of exploration and voyage
that had captured the English imagination long before, in part
because it registered personal responses to landscapes, flora, fauna,
people, and behavior that seemed absolutely foreign to English
experience, and finally because so much of it is beautifully written.
It is, whatever else, *written*, not some pedestrian slogging through
exotic facts to cash in on current literary fads. Beyond that, and for
the purposes of this chapter, it teaches us a lot about how Darwin
became the writer of the *Origin*. *The Journal of Researches* is

Darwin's book of apprenticeship in at least two crafts, scientist and writer. It shows us how, for him, those two vocations were indissoluble.

When Darwin set out on the *Beagle* in 1831, he was not an evolutionist; by the time he came back he was at least inching in that direction; and by the time he published separately the first version of his narrative (1839), he had assuredly become one (largely in secret).[2] And then, by the time he published the second, revised, edition (1845) he had already also written two drafts of the material that was to become *On the Origin of Species*.[3] His experience on the *Beagle* certainly helped trigger his transformation from the orthodox young naturalist of whom the crew made fun because he cited the Bible as conclusive on moral issues, to the evolutionist for whom the Christian narrative simply couldn't be true; but it did so not only because of the facts Darwin unearthed in the course of that five-year trip around the world. The *way* Darwin was learning to see those facts and the *way* he was coming to describe them were at least equally important. The writing itself was part of the process.

Recognizing his deeply felt, very personal, and passionately precise engagement with nature, we ought to be able to get a better feel not only for what he ultimately said but also for the emotional, moral, and aesthetic implications of what he said. The writing of the *Journal*, though different in many respects from that of the *Origin*, anticipates the latter's way of arguing and of representing nature. Looking at aspects of his globe-circling apprenticeship in science, I won't linger on Darwin's early geological walking trips with his Cambridge friend and mentor, John Henslow, or investigate what he already knew by 1831 (which was a great deal). It's clear that even had he become a clergyman (then considered an option), Darwin would have been one of those clergy whom we encounter in Victorian novels, rather more interested in butterfly, or beetle, or mollusk collections than in theological problems or the spiritual condition of his parishioners. The intensity of his instinct for natural history is part of what drove him to become an "author," and was intricately involved in his developing powers of learning to see. At the root of Darwin's success as scientist and author is just that power of close, precise, and extensive observation.

As Darwin had said to his sister, writing was much easier for him when he was simply describing what he saw and experienced, and in *The Journal of Researches*, Darwin could indulge his Romantic longings and relax into description. Descriptive writing does not automatically turn into argument, and certainly not his "one long argument," as Darwin describes the *Origin* to be. On the surface, then, it would seem that looking for Darwin's particular rhetorical and artistic skills budding in *The Journal of Researches* would be misguided: one would seem to miss some of the most characteristic elements of his later prose—for example, the extraordinarily careful connections made among disparate facts in the midst of careful amassing of evidence, and the imaginative thought experiments that require great leaps beyond what is visibly in the evidence, and the minutely particular assemblage of conditional clauses building to strong affirmative conclusions. Amassing of facts there was aplenty on the *Beagle*, but the connections and the logic in the "hodge podge" would seem to be another matter. The *Journal of Researches* is largely a narrative, or set of narratives, controlled usually by chronology or place. But it doesn't take long in the reading to realize that it is certainly not only that.

As against this form, from the time Darwin dared (and was forced) to publish his theory of descent by modification through natural selection, each of his books became a chapter in the one long argument begun publically in the *Origin*, even if, as we read through them, we might miss the argumentative work being done because of the extraordinary and often exuberant profusion of detail Darwin always felt impelled to provide. It is important to recognize that when Darwin is making his long argument, he cannot afford to allow any fact to slip out of the net of natural selection so as to appear incompatible with its workings. In the first chapter, we have seen how all those facts about oddly named ants are pulled together as evidence for the working of natural selection. In fact, the very discussion of those ants is aimed at overcoming a problem that they seemed to pose for the theory. In post-*Origin* writing, every fact requires not only description but a lot of logical and inferential work. No doubt, then, the feel of the prose of the *Journal of Researches* is freer than that of the *Origin*. It breathes some of the air of the *Beagle* and of the wild places Darwin visited. It is full of the wonder of the tropics and of enormous mountains and of exotic species, and in shaping the narrative according to the journey of the *Beagle*, it doesn't always connect the facts. All the connections (by and large) are local. But in what follows I want to suggest that the qualities of meticulous observation, combined with imagination, logical inference, and analogical thinking, so important in developing the case for natural selection, are all on display, often brilliantly, in the *Journal of Researches*.

In the *Beagle* narratives there are moments of intense and widely recognized significance for his later work. In particular, there is Darwin's famous shocked encounter with the Fuegians, whose "primitive" conditions disgusted him. One of the critical moments in Western literature (I'm not exaggerating) is the moment when Darwin records both his disgust and his almost stunned recognition that those wild, naked, primitive creatures were humans: "Viewing such men, one can hardly make one's self believe that they are fellow-creatures."[4] But in fact, though distancing the civilized from the savage as far as he could, Darwin *did* recognize them as "fellow creatures." This episode, and those dealing with Darwin's interactions with the various peoples, and imperial administrators, and genocidal generals, have been widely discussed. No doubt, these interactions are of the greatest importance. But in focusing on Darwin's development as a writer, I find it most useful here to look at some of the less obviously dramatic passages; my emphasis will be on the ways in which Darwin can infuse even the most apparently banal of facts with feelings and significances that put the whole world in motion and turn it into a place of wonder.

As almost all discussions of Darwin emphasize, when he set out on the voyage, he carried with him the first volume of Charles Lyell's *Principles of Geology*, which became a kind of education to him and opened the way to the sorts of imaginative and speculative relations with his subjects that characterize his work in the *Origin*. Lyell helped Darwin learn how to see, not only rocks, but all natural phenomena. The geology, as it introduced forcefully the conception of geological time, infused all rocks and mountains and earth with enormous histories, and taught Darwin how to read the traces of the past, became in his writing—permeated already with Romantic poetry and thinking—revelatory. Darwin's geology in the *Journal of Researches* is as exciting as his biology and anthropology.

It is in the *Journal of Researches*, then, that one can detect a young Darwin developing the powers (along with the facts) that will allow for the great, if occasionally awkward or stumbling prose of the *Origin*. The prose of the *Journal of Researches* does have its "prosaic" moments, to be sure. Darwin was committed to getting all the facts down and thus he often writes as though dutifully ticking off subjects to be covered, the fauna, flora, rivers, and local customs, following the lead of his great model, von Humboldt. But such facts, registered with the sensibility of a young man discovering the world, emerge often in brilliant, intricate, and carefully woven arguments that mix observation and speculation and teach us how Darwin learned to see. The book survives as it does not only because of the intrinsic fascination of the travels it recounts, but because it teaches *us* how to see.

As Darwin looked with enthusiastic attention at the variety of forms of life and matter that he encountered on his five-year voyage and tried to figure out the relations among things, their history, and how he felt about them, he teaches us how to live with the world he so assiduously interprets and describes; even without and prior to his development of the theory of descent with modification by natural selection, his prose implies, in its scrupulous attentiveness, respect for all living creatures, even those that also disgust. Just as Victorian realist fiction was in part an effort to democratize experience, so Darwin's relentless engagement with all kinds of creatures, from the most exotic to the most banal, implies a kind of democratic commitment to the ordinary. In teaching us how to see, it teaches us as well how to make our peace with so difficult a world. It was just because Darwin took nothing for granted in what he saw, probed, and questioned, recorded and reflected, that he was able to move forward from the details to the world-historical theory of descent by modification by natural selection.

A famous passage in John Ruskin's *Modern Painters* captures, if with ironic implications, the point I am after. "Ironic" because Ruskin was appalled by Darwin's ideas and by his science, and thought that Darwin really couldn't "see" in the sense of the word that mattered:

The greatest thing a human soul ever does in this world is to see something, and tell what it saw in a plain way. Hundreds of people can talk for one who can think, but thousands can think for one who can see. To see clearly is poetry, prophecy and religion—all in one.[5]

Ruskin was right, but I would add "science" to "poetry, prophecy, and religion." Seeing entailed for Ruskin the full activity of the mind, the act of perception, *and* the work of the imagination, historical memory, and consecutive thought; but that was true for Darwin too. "Seeing," in this larger sense (and I will be cheating here by including at times hearing or any kind of perception), entails a knowing relationship to the world around us, and a deep feeling for its value and "meaning." The very act of sustained and penetrating attention and, as the etymology of the word suggests, speaking what one sees, make a kind of prophecy. The virtues of "poetry, prophecy, and religion" are the virtues of Darwin's science, as well: a scrupulous, precise attention to particulars of every scale, an intense exploration of their significance and their connections, a refusal to take anything for granted but to find in any the most trivial detail a world of meaning. One of the key elements of Darwin's art is just this capacity to make every detail alive with (at least usually historical) significance.

Darwin not only had to learn to see; he had to learn how to tell what he saw in a plain way. When he complains in his *Autobiography* about his difficulty "in expressing myself clearly and concisely," he notes a "compensating advantage": being forced "to think long and intently about every sentence" (p. 136–7). The sentences show it, even if he never thought of himself as a poet. In the end, his science, and how he learned to tell (and understand) what he saw, had just the virtues Ruskin implies. Seeing is no simple matter; it is not—in the Ruskinian sense—passive. One needs, as Ruskin's *Modern Painters* demonstrates, to *learn* to see, and once having learned, one sees nothing in the same way; everything will speak volumes. How Ruskin saw turned natural objects into moving narratives; how Darwin saw made narratives of rocks and stones and worms and ants.[6] For both, the past is always, however subtly, *visibly* enwound with the present.

But it wasn't Ruskin who taught Darwin how to see, it was Lyell. This, of course, is not news, but in trying to imagine how Darwin, going greenly out to sea, managed to turn a young gentleman's excitement with natural history and some little experience into a world-historical argument, one needs to attend directly to Lyell's great work, *Principles of Geology.* Even more overtly than Darwin's, Lyell's was a grand exercise in teaching how to see. His opening chapters are a tour de force of proselytizing for another way to see the world. He imagines, for example, what the world would seem like to us if we lived as and could see from the point of view of underwater creatures—we would, he suggests, be more likely "to arrive at sound theoretical opinions in geology." Or if we were "entirely confined to the nether world . . . like, some "dusky melancholy sprite," like Umbriel, we would likely "frame theories the exact converse of those usually adopted by human philosophers."[7] What you think depends on what you see, and what you see depends on where you are. Seeing is always, as Ruskin himself showed, incomplete and needs a supplement. Considering what it is we see from different distances, Ruskin concludes, "nature is never distinct and never vacant, she is always mysterious, but always abundant; you always see something, but you never see all."[8] As a good geologist, Lyell knew that and knew that you had to take that into account, to reason as you look, to imagine beyond what you can see. Ruskin, however, distrusted scientific attempts to read beneath the physical surface to physical realities not visible there; he was disgusted by anatomical analysis, and found it prurient, or worse. And thus, although he always saw objects as laden with moral or aesthetic significance, he believed that the "reality" of it was accessible to the naked human eye. The world was to be all surfaces—all that finally mattered is what could be seen. But for Lyell, the geologist, it is just what limited human perspective obscures that needs to be discovered.

His famous insistence that all natural phenomena can be explained by causes now in operation, his uniformitarianism, or what is sometimes called "actualism," requires recognition of what those causes can do and are known to have done, and how those consequences can be read into natural objects now. The new geology that Lyell was trying to create

consists in an earnest and patient endeavour to reconcile the former indications of change with the evidence of gradual mutations now in progress; restricting us, in the first instance, to known causes, and then speculating on those which may be in activity in regions inaccessible to us. It seeks an interpretation of geological monuments by comparing the changes of which they give evidence with the vicissitudes now in progress, or *which may be* in progress. (vol. 3, ch. 1, p. 3).

A huge mountain *might* have been thrown up in one catastrophic moment. But we know—and Lyell carefully measured it—that earthquakes and volcanoes can raise the ground only a few feet; multiply that few feet by several billion years of earthquakes—presto, a thirteen-thousand foot mountain peak. Seeing is patient attention to details; it is comparison; it is inference. It is imagination.

For Ruskin, the seer is a painter or poet. For Lyell he is a good scientist. So too it was for Darwin, yet reading him with attention to the language that produces the ideas everyone cares about, one finds the painter or poet as well, doing other and prophetic work. The art of Darwin's seeing leads to a larger point: while the conventional view is that Darwin's work leads inevitably to a demoralizing and disenchanting vision of both nature in general and humanity in particular, attending to the way Darwin said what he saw in a plain way opens a new and expansive sense of beauty and value. There is a great difference between the Darwin of the *Origin* and the Darwin who is invoked in the name of modern evolutionary biology. Not that modern biology has Darwin wrong, but it is based on certain of Darwin's ideas, and not at all on Darwin the writer, whose books suggest a vision of the world that includes much not particularly important to modern science, as any great book will be more than its ideas.

Darwin's art implies a relation to the world that might be taken as valuable in itself. Is it "nature red in tooth and claw," *or* is there "grandeur in this view of life"? While warily recognizing the former as part of any moral, spiritual, or aesthetic pact we humans can arrange with the secular world as it is, I vote for the latter. The war of all against all is part of Darwin's vision, but it is only *a part* of it, and can be understood fully only in relation to the whole of the *Origin*, to the texture of its language. I choose the "grandeur" because that choice can be read not merely in Darwin's explicit conclusion to the *Origin* but in the writing that cannot be skimmed off the ideas.

II

Early in his voyage on the *Beagle*, Darwin describes his feelings after his first encounter with the "glory of tropical vegetation." "It has been for me a glorious day, giving to the blind man eyes."[9] Despite what he regarded as his wasted years at Edinburgh and Cambridge, Darwin was not even metaphorically "blind" before he set sail, having, among other things, begun an informal kind of naturalist apprenticeship with his teacher at Cambridge, John Henslow. It is clear from his correspondence that well before the directions of his own career were settled and before the invitation to the *Beagle* had been delivered to him, he was busy with insects, microscopes, and aspirations to geology.[10]

What is most astonishing, looking back on the work the voyage produced, is how intensely, thoroughly, meticulously Darwin tried to say what he saw in a plain way; there is something Ruskinianly prophetic in the work, though it was done in the spirit of scientific ambition and rarely aspires to high rhetoric (even where the constant model of Alexander von Humboldt's narrative pushed him, as Darwin's sister complained, to some rhetorical flights).[11] Despite all his efforts, Darwin notoriously missed some very important things—as, in particular, on the Galapagos, the distinctions among the tortoises and the geographical location of the various birds. These birds, now perhaps ironically, were called "Darwin's finches" because, long after he first saw them, they became an important part of the case for his theory. Nevertheless, it is difficult, discovering the sheer volume and quality of the notes and journals, not to be awed by Darwin's relentless questioning of what he perceived, by the perceptiveness itself, combined with his dogged hard work, and by his determination to get it *all* down, and to get it right.

Most of us know the 1845 version of *The Voyage of the Beagle*, the longer, 1839 edition being more difficult to find; not so many of us know (it was not published in his time) his *Beagle* Diary, written perhaps less "scientifically," but rich with detail and human interest. And only Darwin specialists know most of his notes taken on the trip, his *Zoology Notes and Specimens Lists from the H. M.S Beagle*, recently published and edited by R. D. Keynes, or his "Diary of observations on the geology of the places visited during the voyage," and his "Notes on the geology of the places visited during the voyage." Of course, much of the material from the notes becomes part of, and is developed in, the 1845 edition of the *Journal of Researches,* but the volume of notes, even taking that into account, is enormous.

There was, quite literally, a whole world of possibilities to be fished, hunted, watched, and described. What must it have been like, in undertaking such a journey, even to decide how to search for specimens, and then which specimens to collect? What to look for? What to think about? His charge was general and his experience meager. Here is how he saw the problem:

I am often afraid that I shall be quite overwhelmed with the numbers of subjects which I ought to take into hand. It is difficult to mark out any plan & without method on ship-board I am sure little will be done.—The principal objects are 1st, collecting observing & reading in all branches of Natural history that I possibly can manage. Observations in Meteorology.—French & Spanish, Mathematics, & a little Classics, perhaps not more than Greek Testament on Sundays. I hope generally to have some one English book in hand for my amusement, exclusive of the above mentioned branches.—If I have not energy enough to make myself steadily industrious during the voyage, how great & uncommon an opportunity of improving myself shall I throw away.[12]

Somehow, through all this, Darwin managed to keep focused—in part because of his extraordinary passion for work, his apparent love of the things he found, his capacity to read everything, and make meticulous notes, and pursue every potential line of inquiry. The overwhelming nature of his task, so openly defined, helps explain how important Lyell's book came to be for him. Lyell's counter-intuitive reading of geological phenomena, the most extreme of which he continued to try to explain in terms of ordinary causes, made Darwin aware from the start that every object was also a question. For that reason, despite the enormous quantity of specimens Darwin collected and shipped home, he was never merely a collector. Still, collecting and observing minutely was the first step, and it is a revelation to find page after page of meticulous description. I take here randomly a representative sample:

Planaria. Length 1 & 1 & ½ inches. Breadth. 4: oblong: very flat, an elevated line running down the back, sending off lines on each side: Beneath the bands of a yellow substance bordering a central transparent space.—Signs of an aperture at each extremity...

This is not the kind of prose to bring goose bumps to the flesh, but it indicates something of the kind of attentiveness and care—and to the smallest of specimens—that went into his grand compilations of facts. And there are moments when that meticulous observance in his writing rises to something like poetry. We might juxtapose the note about planaria to another mode of attention that allows Darwin, just by virtue of his vigilance, to say what he saw in a plain way, to explain unusual phenomena, and at the same time demonstrate his powers as a writer. During the voyage, for example, he was struck often by finding live insects swimming in the open ocean, and speculates about how that happens. He turns his attention, on November 1st, 1832, to "vast numbers of spiders" on board the *Beagle* more than sixty miles from land. They were "about one tenth of an inch in length and of a dusky red colour." Despite such vivid particularities, Darwin tells the reader that he will refrain from scientific description of the spider, although he does note that "it does not appear to be included in Latreille's genera" (this should serve as a reminder to readers that he always tried to set his observations in context, that he had reference books on the *Beagle* to which to refer, and that his eye was never quite "innocent"). The spiders, in a way that was to be characteristic for Darwin, take on almost human character and behave almost as though they were little Victorians:

The little aeronaut as soon as it arrived on board was very active, running about, sometimes letting itself fall, and then reascending the same thread; sometimes employing itself in making a small and very irregular mesh in the corners between the ropes. It could run with facility on the surface of the water. When disturbed it lifted up its front legs, in the attitude of attention. On its first arrival it appeared very thirsty, and with exerted maxillae drank eagerly of drops of water... Its stock of web seemed inexhaustible. While watching some that were suspended by a single thread, I several times observed that the slightest breath of air bore them away out of sight, in a horizontal line. On another occasion (25th) under similar circumstances, I repeatedly observed the same kind of small spider either when placed or having crawled on some little eminence, elevate its abdomen, send forth a thread, and then sail away horizontally, but with a rapidity which was quite unaccountable. I thought I could perceive that the spider, before performing the above preparatory steps, connected its legs together with the most delicate threads, but I am not sure whether this observation was correct.

All of this, in its businesslike way, is breathtaking. Underlying it all is an assumption that such detailed attention to tenth-of-an-inch creatures is worth it, and will be of interest to readers. The spider becomes a heroic "aeronaut," and the very scale of value in Darwin's world becomes astonishing. Consider the nature of the attention required to make these observations. Who could have noticed the movement of a tenth-of-an-inch spider raising its abdomen, for example? The spiders are not little bugs, they are individuals, whose slightest actions are to be observed and reported. Consider the affective implications of this kind of attention. The spider becomes a kind of heroic character, with a will and intelligence that determine its behavior and make it—an unlikely protagonist in anyone's novel—just that, a protagonist. Consider the prudent caution in the last observation. I am not sure, says Darwin, that the spider actually did "connect its legs together with the most delicate threads." There is a meticulousness and commitment to speaking the truth that is almost as striking as the spider's movements, and an implicit humility as well. But the observation is not merely anthropomorphic and personal. It does work and feels crucial to Darwin because the spider has performed an "unaccountable" act; thus, all of Darwin's attention is turned to explanation. How is this possible? Why does it happen? And then of course, it is difficult not to notice the degree to which Darwin is imagining the consciousness of the little "aeronaut" with the respect he would give to a human. Darwin always tries to *under-stand* the behavior that he sees, and frequently looks for intention-ality. Not only does he recognize (or speculate) that the spider is "thirsty" (when was the last time you thought about spiders being thirsty at all?), but he has to explain why this might be so: "may it not be in consequence of the little insect having passed through a dry and rarefied atmosphere?" The need to explain allows him to see more than anyone else might. Being able to recognize in the first place that something needs explanation entails a quite striking form of attention: the spider lifts its legs "in the attitude of atten-tion," for example. That the spider walks on water inevitably recalls other instances, in colloquialism and in literature, of walk-ing on water. It is a joy to see things in this way, to be able to speculate about whether the spider crosses its legs, for instance. This is to say, the dutifully reportorial passage is almost a celebra-tion of the remarkable powers of that little spider while it implies that extraordinary attentiveness that marks Darwin's entire career. It is evidence if one needed it that, as Darwin says in his *Autobiog-raphy*, he worked to the utmost during the voyage "from the mere pleasure of investigation" (p. 80). Science and art are intricately interwoven.

But that is not all. Darwin goes on in his discussion of spiders, noting that on a later day, he "had a better opportunity of observ-ing some similar facts." So the little narrative just cited is a set of "facts" for Darwin, these descriptions of heroic, disturbed, thirsty, or highly efficient spiders. Here are some more Darwinian *facts*:

A spider which was about three-tenths of an inch in length, and which in its general appearance resembled a Citigrade (therefore quite different from the gossamer), while standing on the summit of a post, darted forth four or five threads from its spinners. These, glittering in the sunshine, might be com-pared to diverging rays of light; they were not, however, straight, but in undulations like films of silk blown by the wind. They were more than a yard in length, and diverged in an ascending direction from the orifices. The spider then suddenly let go its hold of the post, and was quickly borne out of sight. The day was hot and apparently calm; yet under such circumstances, the atmosphere can never be so tranquil as not to affect a vane so delicate as the thread of a spider's web. If during a warm day we look either at the shadow of any object cast on a bank, or over a level plain at a distant landmark, the effect of an ascending current of heated air is almost always evident: such upward currents, it has been remarked, are also shown in the ascent of soap-bubbles, which will not rise in an in-doors room. Hence I think there is not much difficulty in understanding the ascent of the fine lines projected from a spider's spinners, and afterwards of the spider itself.

Note how precision, the explanation of how these insects can be found so far out at sea, the distinctions among species, the analogies with experiences we all will have had, glide into something like poetry—"a vane so delicate as the thread of a spider's web," "like films of silk blown by the wind"—and how the sheer pleasure of the observation is as intense as the explanation itself. The explanatory energy, the assonance and rhythms, and the loveliness of the image are inextricable from each other. We don't, strictly, need the "glittering" in the sunshine, but of course that glitter is not only beautiful. It does work because it allows Darwin to see the form of the threads unmistakably. Moreover, the exotic experience of the spider far out to sea is domesticated by Darwin's appeal to common experience, even to "soap bubbles" that will not rise indoors. The astonishingly rapid movements of these spiders can be understood by analogy with the daily fact of heat waves observable by anyone anywhere. Darwin's prose moves comfortably through the unusual back to the ordinary and along the way intensifies the reader's appreciation of natural phenomena. The more or less anthropomorphized spider becomes, suddenly, attractive, adventurous, and scientifically revealing. But it is hard not to notice the brilliance of the mind doing the perceiving.

If Darwin could do this with spiders, he could do it with rocks. His writings in geology, as Keynes points out, were four times greater than those in natural history; Keynes quotes the autobiography, in which Darwin claims that "the investigation of the geology of all the places visited was far more important, as reasoning here comes into play" (p. ix). In passage after passage, we can watch that reasoning come into play, even in the midst of awed observation of extraordinary phenomena. Raw perception is always tangled with reasoning.

It is no accident that the very first sentence of the first chapter of *The Origin of Species* begins with the phrase, "When we look." In one sense, the whole of the *Origin* is an education in looking, determined by the refusal to take any visible phenomenon for granted, requiring that all things visible be questioned and accounted for, and requiring as well the capacity to observe minutely, detect aberrations, and recognize similarities. Darwin often talks about being "struck" by some object of vision, but he is "struck" because of his uncanny sensitivity to the visual. When he set out on the *Beagle* he did not have a theory on which to base his work, only that remarkable, probably instinctive capacity to "see," to "look," and to be "struck by" natural phenomena.

It is crucial to recall that "seeing" is not passive and that for Darwin it was associated with many other mental activities. Even (perhaps especially) those who argue most strenuously against evolution agree that despite his protests to the contrary, he did not work strictly on the "true Baconian principles" he invoked in his *Autobiography*. One of the remarkable and for me most impressive aspects of his career is that through the years he was *not* shaken from his theory by the possibilities of counter-evidence (although through further editions of the *Origin*, he increasingly allowed for speciation in part through use inheritance, an idea that is, however, present from the start). Even in the face of William Thomson's demonstration that, given the rate of the cooling of the earth, it could not be old enough to have produced by natural selection the organisms now living in it, he persisted. He managed always to contrive his own counter-interpretations, frequently built on extraordinary mind experiments. These were intellectually adventurous exercises of his remarkable imagination—that is, of his capacity to "see" what was not immediately visible, or visible only through traces that ordinary observers might not notice. On the other hand, when he claimed, in the *Origin of Species*, that certain facts, if proved true, would be "fatal to my theory," he was not being disingenuous. Nor was he falsifying when he said that "I have steadily endeavoured to keep my mind free, so as to give up any hypothesis, however much beloved...as soon as facts are shown to be opposed to it" (*Autobiography*, 141). Partly from intense curiosity and partly to be certain that he had the facts for his arguments, he thus busied himself with collecting facts even though, as we have seen, he was "often afraid I shall be quite overwhelmed with the numbers of subjects which I ought to take in hand."

The power to notice things depends upon a strenuous mental alertness (as a bird watcher I can testify that I often have in my field of vision birds that I do not "see" until they move), and on a complex mental condition—the recognition of anomalies, which in turn depends on a very rich knowledge of context and of what is taken for the norm, a gift for analogy, a capacity to recognize connections, and rare interpretative ability. We have seen something of these qualities already in the passage about spiders. But while Darwin never quite claimed for himself exceptional powers of reasoning, he rather awkwardly, against his own instinctive modesty, allowed that he must have had some power of "reasoning" since, as he rightly asserted, "the *Origin of Species* is one long argument from the beginning to the end" (*Autobiography*, 140).

Thus, although his rhetoric and his working style demanded the strictest observance of what the world presented to him, he did not see innocently. "His pleasure came less," Gertrude Himmelfarb argues, "from the passive, sensual act of seeing than from the effort of comprehending and analyzing."[13] He could not, he admits, "resist forming [an hypothesis] on every subject" (*Autobiography*, 141). Himmelfarb errs here only in that "the sensual act of seeing," if not often passive, was intensely pleasurable for Darwin, as passage after passage of *The Voyage of the Beagle* and even of the *Origin* make plain. The sensual pleasure was enwound with the intellectual pleasure of the hypothesis. Direct engagement with the natural world—the exotic tropics and spiders at sea and the weeds in his garden—carried him through his virtually lifelong painful illness and beyond his loss of feeling for the poetry that had inspired and taught him in his early *Beagle* years. "The excitement" [from scientific work], he wrote, "makes me for the time forget, or drives quite away, my daily discomfort" (*Autobiography*, 115). Experiencing and questioning were not distinct for him but part of the entire experience, and his frequent use of the word "wonder" as he describes some impressive landscape or phenomenon, and talks of "wonderful facts,"[14] suggests just that unity of feeling and thinking in Darwin's approach to nature that I am trying to identify in his writing. The word "wonder" for Darwin was not merely a vague exclamation of awe; the awe is felt just because the phenomenon Darwin looks at is so apparently unusual that it immediately raises questions and requires answers.

It is the peculiar combination of "reasoning" with sensual pleasure and observational instinct that gives to Darwin's work, both as scientist and writer, its characteristic quality. Michael Ghiselin and many others who followed him have rightly emphasized Darwin's "hypothetico-deductive" method.[15] But while Darwin's imagination was fertile, his hypotheses came to him from intense (often visual) experience that preceded them. Comparison was a critical tool of perception, essential even to defining correctly what was being observed: one tells the size of something by comparing it to something else one knows better; one recognizes relationship by recalling another creature that has similar characteristics. Obviously then, since Darwin was always comparing things in order to see accurately, analogy became a critical figure in his prose. Certainly, the hypothesis of "natural selection" came to Darwin only after long years on the *Beagle*, only after recognizing relationships through comparison, and then finding a coherent way to account for the spectacularly beautiful, abundant, and diverse worlds that he "saw" in his travels.

That famous flash that he claims followed his re-reading of Malthus in 1838 was well prepared for. That endless series of questions that each observation provoked, elicited at each moment during his *Beagle* voyage, a local answer, but those local answers began to take on a recognizable pattern, as we will see, a pattern that prepared the way for the larger theory. He prodded himself into those questions: why does that animal look like other, different animals encountered elsewhere? Why is it here and something similar there? How did it get here? How does it nourish itself? How does it propagate? Who are its enemies? Why are its organs arranged as they are? Why is it of this color rather than that? What past phenomena contributed to its current conditions? Everything required explanation. The raw sight and the questions were instantaneously and cumulatively linked, and as those raw materials were juxtaposed and rethought, it became clear that new explanations would be necessary."The limits of man's knowledge in any subject," he wrote, "possess a high interest, which is perhaps increased by its close neighbourhood to the realms of imagination" (*Beagle*, 285).

What Darwin saw in that first thrilling tropical exuberance became part of a vast storage house of experience that allowed the blind man to see and extended the visible world into the not obviously visible, or visible only with the supplement of imagination and reason. Seeing with Darwin's eyes meant filling the world with meaning, registering all things visible in relation to other things and other times. Seeing with Darwin's eyes entails active imaginative intelligence that can pry from the visible the traces of an invisible past and of minutely suggested unobvious relations. Such seeing gives meaning to the most ordinary of natural objects, as it also, therefore, profoundly shapes our sense of value.

III

In his otherwise almost awkwardly modest *Autobiography* Darwin claims to think "that I am superior to the common run of men in noticing things which easily escape attention, and in observing them carefully" (141). This superiority is evident already in a myriad of details presented in his *Journal of Researches*. He had realized during the voyage that "the pleasure of observing and reasoning was a much higher one than that of skill and sport" (79) although he had been an avid hunter and shooter all through his early youth. And he developed on the voyage "the habit of energetic industry and concentrated attention" (79). In what follows I will be looking at some selected passages in the *Journal*, primarily about geological phenomena, that reveal Darwin learning to see and to say what he saw in a plain way. The prose becomes the vehicle for his vision, blending precision of perception with intensity of feeling. I choose these passages in part because they are not ostensibly about "evolution," and because they do not engage with those large issues of human development and human relation to animals that seem to require some larger theory. But in the very pattern of their perceptions and their development, they do seem to me to anticipate the larger arguments for which Darwin is famous, and they demonstrate clearly how Darwin developed that way of seeing that led to those arguments. They are very clear examples of how "observing and reasoning" were for him part of the same experience. The world of traces that Darwin eyes reveal "speak" of other times, of other things, of transformations; stones, rocks, and clouds, imposing in themselves, are recognized immediately as signs. Observing and reasoning together overturn the implications of the simple passive seeing with which we all tend to move through the world. Throughout the *Journal* virtually everything seen reverberates with a "meaning" beyond what is immediately visible, and with that feeling of wonder that is the persistent mark of Darwin's observations of the world, down to his last days.

I want to emphasize several patterns of seeing that emerge consistently from his descriptions: the most static of things observed is seen as part of a world in the process of constant, if not always visible, change; every space is filled down to its most obscure niches with teeming life, entirely and complexly connected (and entangled); everything means something beyond itself, something not visible but traceable. Everything, even the most grotesque or hideous, is fascinating.

Perhaps the most important aspect of his way of seeing is his persistent visualization, through careful perception of things present, of things *not* present. We can find a model for this quality in his discussion of his experience in Bahia Blanca, in which he talks of how the natives attend to the "rastro," or track, that fugitives leave. And he notes that

One glance at the rastro tells these people a whole history. Supposing they examine the track of a thousand horses, they will soon guess the number of mounted ones by seeing how many have cantered; by the depth of the other impressions, whether any horses were loaded with cargoes; by the regularity of the footsteps, how far tired; by the manner in which the food has been cooked, whether the pursued travelled in haste; by the general appearance, how long it has been since they passed. (101–2)

This is both rhythmically impressive and an excellent representation of the way Darwin learned to look at the world. The passage depends in part on an emotion not stated but implicit in its rhetoric, astonishment at the natives' capacity to read, "at one glance," important meaning into visual evidence. It is an astonishment of the sort that the first readers must have felt from Darwin's descriptions of the tropics, mountains, islands, and volcanoes that so awed him, and that at the same time he managed to interpret with such confidence. Implicit here is the natives' uncannily developed perceptiveness and intelligence, their power to create a logical and, indeed, useful sequence of thought out of footprints. Such a way of seeing the slightest details turns the world into a set of traces, but not merely of antiquarian value, for those traces reveal things profoundly important to the way the natives act at the moment. Throughout the voyage, Darwin reads the "rastro" of nature itself, and through his readings the world takes on meaning everywhere. The quality is not, of course, only Darwinian: Cuvier and Richard Owen were famous for being able to reconstruct fossil bones into integral structures that could render dinosaurs real, and make them part of contemporary culture. Sherlock Holmes and the detective novel await.

The effect of reading the traces of the world is the production of narratives, of stories; and Darwin's theory, as he was to develop it, like his writing on the *Beagle* voyage, becomes a long series of stories. Later in the *Beagle* narrative, he notes how as he explored in the wilderness, he always felt "a strong desire...to ascertain whether any human being has previously visited an unfrequented spot. A bit of wood with a nail in it, is picked up and studied as if it were covered with hieroglyphics." The world is a written story, and one needs only the experience and power to read the language. So,

Possessed with this feeling, I was much interested by finding, on a wild part of the coast, a bed made of grass beneath a ledge of rock. Close by it, there had been a fire, and the man had used an axe. The fire, bed, and situation showed the dexterity of an Indian; but he could scarcely have been an Indian, for the race is in this part extinct, owing to the Catholic desire of making at one blow Christians and Slaves. I had at the same time some misgivings that the solitary man who had made his bed on this wild spot, must have been some poor shipwrecked sailor, who, in trying to travel up the coast, had here laid himself down for the dreary night.

Darwin can make some ultimately complicated inferences about the "man" who had been there. Although he doesn't manage to solve the mystery, he does succeed in telling a story in some detail—even adding to the evidence of fire and axe deductions from major and indeed appalling historical events—about who the person was and how he got there. Darwin "reads" the world as though it were written in "hieroglyphics" and in so doing, he fills everything in nature with significance. Ironically, Darwin, accused of draining the world of meaning, in fact fills it with meaning. Of course, "meaning" here doesn't signify answers to the great spiritual questions, but it does mean that every detail of one's life, every detail of the natural world is like writing—it signifies, telling us about history, relationships, and future possibilities.

Since Darwin is dealing here not with rocks but with humans, it is not surprising that the discovery turns into tentative stories. But here it is easier to detect the feeling that attaches to storytelling than when Darwin traces the history of rocks, as he sympathetically registers the shipwrecked sailor's plight and the "dreary night" he must have experienced (283). Nevertheless, affect clings around other less humanly oriented stories. Argument and narrative are one—scientist and storyteller, scientist and prophet are incipient even in so minimal a narrative as this. Darwin's world cries out to make sense.

But while the Nature Darwin aspires would be entirely interpretable, it is at the same time full of wonders and mysteries. There may be no "mysterious, incalculable forces" in Darwin's world (the phrase is Max Weber's in his essay claiming that modern science has "disenchanted" the world[16]), but the "calculation," if that is the right word, is both intellectually and emotionally exciting, and the recognition that everything in that world, no matter how minuscule, issues out into enormously complex and rich sets of relations is for him and for attentive readers as wonderful as the mysterious, incalculable forces immanent in the pre-scientific world. Whatever inferences might be drawn from the theory of descent by modification through natural selection, Darwin's prose—his registration of what he sees, what he imagines, and the significance of the rastro—produces over and over again a sense of wonder.

Here is one example of how, while each perception seeks its meaning, the science and the spontaneity of his response to nature combine in what I would also call poetic insight. He writes of the "din of rushing water" in a river in the Cordilleras, and of how the noise from the "the thousands and thousands" of stones, as they "rattled one over another" was heard night and day "along the whole course of the torrent" and "spoke eloquently to the geologist" as, with "one dull uniform sound," they all hurried "in one direction." (ch. 15, p. 318). Here is a perfect example of the "double movement" of his prose, the movement that achieves its rational triumphs by evoking the wonder that makes the best argument *against* the idea that Darwin's work implies a very bleak vision of the world:

It is not possible for the mind to comprehend, except by a slow process, any effect which is produced by a cause repeated so often, that the multiplier itself conveys an idea, not more definite than the savage implies when he points to the hairs of his head. As often as I have seen beds of mud, sand, and shingle, accumulated to the thickness of many thousand feet, I have felt inclined to exclaim that causes, such as the present rivers and the present beaches, could never have ground down and produced such masses. But on the other hand, when listening to the rattling noise of these torrents, and calling to mind that whole races of animals have passed away from the face of the earth, and that during this whole period, night and day, these stones have gone rattling onwards in their course, I have thought to myself, can any mountains, any continent, withstand such waste?

Or withstand such a long, Ruskinian sentence? Even the rhythms of this prose conspire with the vision to create that remarkable sense of vast stretches of time flowing inexorably toward enormous geological transformations. The Darwinian gaze connects immediately the "mud, sand, and shingle" of the beaches with the insistent "rattling noise" of those mountain torrents. In just this way, by recognizing connections, Darwin transforms the quotidian, the habitual, the trivially normal—worms, ants, bees, pebbles—into the *almost* incomprehensibly vast and uncontainable. As Darwin sees and hears the little stones, he cannot help also recognizing their history and inferring their future. He cannot resist the implications for the mud many thousand feet deep, nor for what that means about the extent of time of which—as he connects the two—this landscape gives evidence. After such a passage no peaceful gurgling and rattling stream will be the same. It is a passage of Lyellian uniformitarian geology, but it does other kinds of work as it makes the extraordinary comprehensible exclusively in terms of the working of ordinary nature. What Darwin *sees* immediately seems incomprehensible; his prose implies a force beyond nature, and as narrator Darwin puts himself in the place of a reader who has to be skeptical. But as narrator he insists on a grand historical perspective, which entails an act of imagination, and the visible is thus translated; the geological phenomena become naturalistically comprehensible, but with a sweep of feeling that reflects awe at the power of this ordinary. The phenomena are made comprehensible, but only in a moving, sweeping revelation of the power of these ordinary forces moving through time.

This is the double movement that emerges everywhere as Darwin seeks to understand the traces that the visible world leaves of its past—inferring history, meaning, movement; explaining phenomena that many of us would not think needed explanation. That movement doesn't always work out through an initial expression of awe. Here is another example, this one very quietly understated. Darwin notes how a quartz formation on the top of a large mountain abutting a "sea-like plain" gives evidence of the working of waters no longer visible on the largely arid cliffs:

I do not think that Nature ever made a more solitary, desolate pile of rock;—it well deserves its name of *Hurtado*, or separated. . . . The strange aspect of this mountain is contrasted by the sea-like plain, which not only abuts against its steep side, but likewise separates the parallel ranges. The uniformity of the colouring gives an extreme quietness to the view;—the whitish grey of the quartz rock, and the light brown of the withered grass of the plain, being unrelieved by any brighter tint. From custom, one expects to see in the neighbourhood of a lofty and bold mountain, a broken country strewed over with huge fragments. Here nature shows that the last movement before the bed of the sea is changed into dry land may sometimes be one of tranquility. (109)

It is a quiet evocative description, the very texture of the stone registering the "tranquility" of the death of the waters that had to

that point been wearing down "the pile of rock." No great catastrophe created this scene.

But Darwin's curiosity, once he has made his inferences about
the area's geological history by working out the significance of this
"rastro," emerges yet again—as though to prove the story he
already believes: "under these circumstances I was curious to
observe how far from the parent rock any pebbles could be
found. On the shores of Bahia Bianca, and near the settlement,
there were some of quartz which certainly must have come from
this source: the distance is forty-five miles." (ch. 6, p. 109).

In a way, this passage is more characteristically Darwinian than the
passage about the rattling stream, for this one is understated—almost
as though to match the apparently rather quiet history of this mountain. He announces his curiosity and satisfies it within a few words:
"the distance is forty-five miles." And yet, of course, this is astonishing, as awesome as the power of the rattling stream that wears away
mountains. In fact, it implies, without bothering to assert it, just that
kind of power of water that, this time, has long ago dried up. It all fits
neatly into the Lyellian interpretation of this rastro; the geological
phenomenon is explained implicitly by reference only to normal
causes.

Part of the secret of Darwin's power as a writer is understatement. While there are moments in *The Journal of Researches* and
in some of his other writings, in which one can see Darwin
straining, with more or less success, towards what he would have
thought of as Humboldtian rhetoric, the richest and most beautiful
of his writing is, as Adam Gopnik suggests, apparently straightforward, attentive, direct, rather like this little passage of discovery.
To get the full feeling of Darwin's writing one must register how
much is implied in such a quiet representation of his discovery.
Consider how many days of travel it took to get from the peaks to
the beach, how absolutely different the two environments, and
thus how profoundly imaginative the connection between little
pebbles one takes for granted on a beach and the sublime mountains thousands of feet high, days and miles away. The feeling is

not only awe at the idea that the mountain, under the pressure of waters no longer extant has strewn pebbles forty-five miles away, but astonishment that Darwin's power of reading the rastro led him to *look for* those pebbles. The awe at the natural phenomenon is compounded by the admiration of the mind that reads such phenomena so precisely. Such a mind endows the smallest pebble with a history almost as sublime as that of the mountain range.

The force of the story is not only in the history of the natural object, but in the observer's powers to read it, that is, the act of counter-intuitive discovery itself, the imagination that connects large and small through space and time. The reader responds not only to the sublime panorama in time and space but to the activity of the mind: both are startling.

The combination of observation and reasoning marks Darwin's treatment of virtually all geological subjects. It is all done casually enough that one need not notice how much speculation is involved—it is, after all, simply geological observation. But such attention to the movement of the observations and reflections upon them makes the often strange landscapes Darwin comes upon exciting and energizing, and at the same time provokes later seeing (as of the quartz pebbles on the beaches forty-five miles away). There is a long sequence, for example, in which Darwin describes a phenomenon that he and, apparently, virtually all others who had seen it on one of the Falkland Islands, called "streams of stones" (p. 197–9). It entails very precise description of the "stream" and of particular stones, but it is held together by Darwin's intense questioning of the whole phenomenon. The "myriads of great loose angular fragments of the quartz rock," varying in size "from one or two feet in diameter to ten, or even more than twenty times as much" are not, he notes, "thrown together in irregular piles" but "spread out into level sheets or great streams." The angles of the stones are not rounded so that, he infers, water is probably not responsible for this formation. Much of what he sees he cannot see completely, but he also infers the enormous depth of the stones —which cannot be precisely measured—from the sound of water trickling many feet below the surface, and from the further inference that the "crevices between the lower fragments must long ago have been filled up with sand."

He is particularly struck ("the most remarkable circumstance") by their "little inclination," which is often "only just sufficient to be clearly perceived." In a characteristic analogical mode to explain the conditions of this overwhelmingly strange place, he notes that "the slope would not have checked the speed of an English mail coach." The precision of description continues, and turns once again into Darwinian poetry just by being so careful and opening the unfamiliar to the familiar:

a continuous stream of these fragments followed up the course of a valley, and even extended to the very crest of the hill. On these crests huge masses, exceeding in dimensions any small building, seemed to stand arrested in their headlong course: there, also, the curved strata of the archways lay piled on each other, like the ruins of some vast and ancient cathedral. In endeavouring to describe these scenes of violence one is tempted to pass from one simile to another.

Darwin resists much simile making, though the cathedral and even the word "stream" do metaphorical work. He is driven by a need to explain such a strange and awesome phenomenon, which he experiences, in all its raw beauty, as a "scene of violence." And in the rest of the paragraph he works out an explanation, although in the end, he can only invoke an "enormous convulsion" that fractures the landscape into "myriads of fragments."

The convulsion, of course, is not visible, but what Darwin sees implies for him violence so great that he is *almost* overwhelmed into endless similes. In the end, he returns to the phrase "stream of stones," a description that "immediately occurred to everyone." The simile is weaker than the phenomenon, which Darwin re-enacts in his prose. History is right there in the sharp angles and regular slight incline of those enormous stones.

Finding a "great arched fragment, lying on its convex side," he asks whether we must believe "that it was fairly pitched up in the air, and thus turned?" And there follows another characteristic sequence of reasoning, based on imagination, analogy with other phenomena, and probability: "as the fragments in the valleys are neither rounded nor the crevices filled up with sand, we must infer that the period of violence was subsequent to the land having been raised above the waters of the sea." While "in a transverse section of the valley, the bottom is nearly level" and the "fragments appear to have travelled from the head of the valley," it seems more probable, Darwin claims, "that they have been hurled down from the nearest slope; and that since, by a vibratory movement of overwhelming force, the fragments have been leveled into one continuous sheet." He reads the visible present against his experience of the Concepcion earthquake directly into the stream of rocks, and infers violence yet more radical:

what must we say to a movement which has caused fragments many tons in weight, to move onwards like so much sand on a vibrating board, and find their level? I have seen, in the Cordillera of the Andes, the evident marks where stupendous mountains have been broken into pieces like so much thin crust, and the strata thrown on the vertical edges; but never did any scene, like these "streams of stones," so forcibly convey to my mind the idea of a convulsion.... Yet the progress of knowledge will probably some day give a simple explanation of this phenomenon, as it already has of the so long-thought inexplicable transportal of the erratic boulders, which are strewed over the plains of Europe. (198–9)

There again is the double movement of Darwin's prose: the almost breathless description of awe-inspiring phenomena, followed by the confident assertion that it will all be explained naturalistically and simply. Analogy operates as well. Part of Darwin's confidence that it will all be explained simply results from his recognition of the parallel between the "erratic boulders" of places like Stonehenge in England, and these massive, violently placed stones. The simplicity and quasi domestication, however, cannot minimize the intensity of perception, the awesomeness of the phenomenon, the extraordinary explanatory inferences. It is no wonder that the *Journal of Researches* was a popular book.

These kinds of geological investigations, as we can already infer from the examples I've used, provide an intellectual framework for the book in all its versions. The same strategy I have pointed to in the passage about the stream emerges in a more obviously "sublime" moment later in the narrative. Seeing one thing, Darwin calls to mind another, or understands that he will need to find parallels before seeing adequately what is directly in front of him. He notes, for example, how the summit of an Andean mountain "was much shattered and broken into huge angular fragments," but he observed "one remarkable circumstance," that the surface "presented every degree of freshness—some appearing as if broken the day before." The perception is immediately allied to an idea—he "fully believed that this was owing to the frequent earthquakes" (p. 259). Recognizing, however, how easy it was to be deceived on this matter, and thus "doubting its accuracy," he does not confirm it to himself until he sees a similar landscape on the top of Mount Wellington in Van Diemen's land, a place in which now there are no earthquakes, where the shattered broken rock all appeared "as if they had been hurled into their present position thousands of years ago." This simple argument entails a whole set of activities—the act of seeing and imagining how rocks might be so shattered; the immediate inference of the age of the shattering; the later analogy that confirms the inference and the imagination. Analogy here does fine literary work, but is a condition for solid scientific observation. The rocks tell their stories.

And that story continues as, at the summit, Darwin describes himself "taking pleasure from the scenery," viewing from a sublime height virtually all of Chile laid out before him as on a map. But the sheer passive pleasure in the breathtaking view is, as he says, "heightened by the many reflections" arising from a view of the Campana range, the details of which lead him to wonder "at the force which has upheaved these mountains, and even more so at the countless ages which it must have required, to have broken

through, removed, and leveled whole masses of them" (p. 260).
Those reflections depend on Darwin's recalling "the vast shingle
and sedimentary beds of Patagonia," which he recognizes as debris
from these very mountains, and which, he thinks, "heaped on the
Cordilleras, would increase its height by so many thousand feet."

When in that country, I wondered how any mountain-chain could have supplied
such masses, and not have been utterly obliterated. We must not now reverse the
wonder, and doubt whether all-powerful time can grind down mountains—even
the giant Cordillera—into gravel and mud. (p. 260)

"We must not now reverse the wonder." The beautiful vision of the
Campanas becomes utterly sublime, not because its history is
beyond human reflection and calculation, but because what Darwin successfully calculates is so clearly awesome in its extent and
violence. Darwin comes to *see*, by way of juxtaposition of the
shingle and sediment remembered with the already high mountains before him: they had been even more massively higher, and
they were so millions of years before. Wonder piles upon wonder—"we must not reverse it," he urges, "by doubting the extent
and power of time." The monumental landscape, so grand and
solid, turns into matter as fluid as the ocean: "Daily it is forced
home on the mind of the geologist," Darwin writes, "that nothing,
not even the wind that blows, is so unstable as the level of the crust
of the earth" (p. 323). And then, much later on, near the end of the
book, almost as though to summarize the myriad details of his
experience, he asks: "where on the face of this earth can we find a
spot on which close investigation will not discover signs of that
endless cycle of change, to which earth has been, is, and will be
subjected?" (p. 493). Darwin's seeing remakes the visible world by
way of an imagination that can visualize its ancient and future
meanings. What is solid, monumental, and unmoving is, rather, in
constant flux. Seeing, history, and predictive science blend in what
certainly the readers of Darwin's very popular book must already
have felt—an enchanting imagination of the natural world.

Darwin moves us from quartz pebbles to an overwhelming understanding of a world both monumental and in constant flux—the small becomes enormous. The wonder is not reversed. It is still "wonderful," but we know now that the daily little movements of ordinary things, things we experience daily, the trickling of little streams—all these make part of a solidly substantiated grand rhetorical flourish that transforms the stable world we occupy into something like the gossamer spiders' web in the wind: "nothing, not even the wind that blows, is so unstable as the level of the crust of this earth." Taken in its most literal form, this could, of course, lead to bleak inferences—a little *carpe diem* might seem in order. But experience in its breathless and wonderful movement out of slight domestic details does indeed make the world "wonderful." The prose does the work that denies the bleakness it might seem to be affirming.

But while geology certainly gave Darwin, via Lyell, his sense of the flux of things and the vastness of time, a sense that even now it is difficult to imagine (and the failure to imagine it inevitably results in the invocation of transcendental powers to explain the phenomena that Darwin describes here), such juxtapositions of ostensibly unrelated things is part of all the sciences Darwin learned to practice on the *Beagle*.

Consider how he looks at the almost endless varieties of life, which he so sedulously collected, labeled, analyzed, and shipped home. Darwin's seeing leads here also to counter-intuitive ways of understanding not only rocks but organic life itself. One passage must stand in for many, and might be seen as indicative, as well, of how his observations regularly triggered questions, and how reflection and vision were virtually simultaneous. Here he describes the life on a salt lake:

The border of the lake is formed of mud: and in this numerous large crystals of gypsum, some of which are three inches long, lie embedded; whilst on the surface others of sulfate of soda lie scattered about.... The mud is black, and has a fetid odour. I could not at first imagine the cause of this, but I afterwards perceived that the froth which the wind drifted on shore was coloured green, as if by confervae.... Parts of the lake seen from a short distance appeared of a reddish colour, and this perhaps was owing to some infusorial animalcula. The mud in many places was thrown up by numbers of some kind of worm, or annelidous animal. How surprising it is that any creatures should be able to exist in brine, and that they should be crawling among crystals of sulfate of soda and lime! And what becomes of these worms when, during the long summer, the surface is hardened into a solid layer of salt? Flamingoes in considerable numbers inhabit this lake, and breed here; throughout Patagonia, in Northern Chile, and at the Galapagos Islands, I met with these birds wherever there were lakes of brine. I saw them here wading about in search of food—probably for the worms which burrow in the mud; and these latter probably feed on infusoria or confervae. Thus we have a little living world within itself, adapted to these inland lakes of brine. A minute crustaceous animal (Cancer salinus) is said to live in countless numbers in the brine-pans at Lymington; but only in those in which the fluid has attained, from evaporation, considerable strength—namely about a quarter of a pound of salt to a pint of water. Well may we affirm, that every part of the world is habitable! Whether lakes of brine, or those subterranean ones hidden beneath volcanic mountains—warm mineral springs—the wide expanse and depths of the ocean—the upper regions of the atmosphere, and even the surface of perpetual snow—all support organic beings. (65–6)

Here again the characteristic Darwinian moves. Careful attention to details that in themselves are not particularly fascinating; meticulous measurement; but also, immediate questioning: why is the mud black? Why is the froth green? Why does the lake, from a short distance, seem reddish? What happens to the worms when the surface turns entirely to salt? What do the flamingoes find to eat? What do the worms, that the flamingoes eat, themselves eat? Are there analogous situations in his experience? He recalls Lymington and the quantity of salt in its waters. The questions turn the scattered vision of separate phenomena into a connected world, "a little world within itself," self-sustaining in these apparently uninhabitable islands of brine. Suddenly the details explode into a general revelation: "every part of the world is habitable," and Darwin's mind and prose leap across the world, to snow-covered mountains, subterranean lakes, and the upper regions of the atmosphere. This vision of abundance, of life finding a place for itself everywhere, is a step away from Darwin's famous metaphor of the wedge: "The face of Nature may be compared to a yielding surface, with ten thousand sharp wedges packed close together and driven inwards by incessant blows, sometimes one wedge being struck, and then another with greater force" (*Origin*, 67). But the metaphor here would be, I think, somewhat less violent. The whole world is flooded with life; life-forms adapt to the most unlikely places; every niche is filled. The banal and the fetid once again issue in wonder. The strategy of the prose is familiar by now—a strange phenomenon explained, the explanation more exciting than the original strangeness.

Let me take just one more example from the teeming world of Darwin's first important book, an appropriately Darwinian anti-climactic one, characteristically Darwinian in that it is not from the main text, but from a footnote that Darwin appends to a very brief notation of having discovered "a singular little mouse (M. brachiotis)" on several of the islets of the Chronos Archipelago (p. 289). That mouse appeared on some of the islets, but not all, and the appearance evokes in Darwin another fit of wonder: "What a succession of chances, or what changes of level must have been brought into play, thus to spread these small animals throughout this broken archipelago!" Attention to a mouse is the beginning of this note. Mice are not to be ignored in Darwin's world, nor spiders, nor ants, nor worms. Each living thing is important in Darwin's prose; each raises a broad range of questions. Once noticed and thought of, the mouse requires an explanation. How did the mouse get here? What strange assemblage of "chances" could have done it? The mouse gets only a footnote: Darwin starts with causes now in operation as he speculates about the chances. He points out that "some rapacious birds bring their prey to their nests." Invoking again long stretches of time, Darwin suggests that over the course of centuries some of the prey is likely to have escaped. Driven by the logical necessity to explain the mouse's presence, Darwin concludes more severely and forcefully, that "some such agency is *necessary*, to account for the distribution of the smaller gnawing animals on islands not very near each other." Seeing the mice leads Darwin to seeing the rapacious birds, and connecting them because flying is the most likely means of transportation across the archipelago. Knowing the habits of the birds, Darwin can begin to make sense of the erratic presence of the mice.

This is the way Darwin teaches us how to see, in part by making us recognize that one must see beyond the visible, and that every instant of perception is charged with assumptions, and that every perception should lead us to questions, and that every question makes the world both more interesting, and richer. Everything is exciting and valuable in this world of phenomena, questions, analogies, and connections. That such seeing, and telling what is seen in a plain way, emerged so significantly some years later in the *The Origin of Species* is not surprising. There the refusal to take anything for granted, together with Darwin's determination to attend to the minutest organisms, structures, pebbles, fissures, and colors, manifests itself in the grand synthetic argument we all know. One consequence of a close engagement with the prose of *The Journal of Researches* is a recognition that the world, flattened into a generalization about nature red in tooth and claw (I know it's not Darwin's phrase), is utterly inadequate to the experience of the world Darwin offers us through his language. The *Journal* does, however, help explain how, from the smallest of details, Darwin could have moved to his world-historical generalization. For he had already moved us from the brine-dwelling organisms to the regions of perpetual snow, to the upper atmosphere, and to a recognition that life will find its way anywhere. And in so doing, his meticulous and yet imaginative prose presents a world that is too beautiful, too complex, too laden with meaning to justify the usual bleak inferences from the theory to which it would give birth twenty years later.

Notes

1.   In the title of the second edition, "Geology and Natural History" were inverted.

2.   See Niles Eldredge, *Darwin: Discovering the Tree of Life* (New York: W. W. Norton, 2005) and Philip R. Sloan, "The Making of a Philosophical Naturalist," in *The Cambridge Companion to Darwin*. The first time he let anyone outside of his family know his views was in a letter, to Joseph Hooker, the distinguished botanist and perhaps Darwin's closest friend, which included the now famous line, "it was like confessing a murder!"

3.   The two drafts, a short one written in 1842 and a short book-length one in 1844, are easily available in the edition edited by Francis Darwin, *The Foundations of the Origin of Species* (BiblioBazaar, 2008).

4.   For an extended discussion of the significance of Darwin's frequent recurrence to this experience, see Cannon Schmitt, *Darwin and the Memory of the Human: Evolution, Savages, and South America* (Cambridge: Cambridge University Press, 2009).

5.   John Ruskin, *Modern Painters*, III, sect. 28, ch. 16, p. 333 of the fifth volume of *The Works of John Ruskin*, eds. E. T. Cook and Alexander Wedderburn (London: George Allen, 1904).

6.   For a discussion of the curious overlap of interests and powers in two writers so fundamentally antithetical, see my essay, "Ruskin and Darwin and the Matter of Matter," in *Realism, Ethics, and Secularism: Essays on Victorian Literature and Science* (Cambridge: Cambridge University Press, 2008).

7.   Charles Lyell, *Principles of Geology* (Chicago: University of Chicago Press, [1830] 1990), 82–3.

8.   *Modern Painters*, I, sect. 2, ch. 5, p. 329.

9.   *Charles Darwin's "Beagle" Diary*, ed. R. D. Keynes (Cambridge: Cambridge University Press, [1988] 2001), 23.

10.   Jestingly, a friend writes Darwin in the spring of 1831, "have you bottled any more beetles, or impaled any butterflies?" *CCD*, From Henry Matthew, vol. 1, March or April 1831, 119. Complaining about the hard work entailed in preparing for exams, Darwin notes to his good friend W. D. Fox, "I actually have not stuck a beetle all this term" (111). His *Beagle* letters are punctuated with notations about insects received and sent.

11.   The whole paragraph of Caroline's response to the journal is worth quoting because it not only criticizes too much flourish but nicely reflects on the nature of Darwin's best prose:

> Your writing at the time gives such reality to your descriptions & brings every little incident before one with a force that no after account could do. I am very doubtful whether it is not *pert* in me to criticize, using merely my own judgment, for no one else of the family have yet read this last part—but I *will* say just what I think—I mean as to your style. I thought in the first part (of this last journal) that you had, probably from reading so much of Humboldt, got his phraseology, & occasionally made use of the kind of flowery french expressions which he uses, instead of your own simple straight forward & far more agreeable style. I have no doubt you have without perceiving it got to embody your ideas in his poetical language & from his being a foreigner it does not sound unnatural in him—Remember, this criticism only applies to parts of your journal, the greatest part I liked exceedingly & could find no fault, & all of it I had the greatest pleasure in reading. (*CCD*, I, 28 October, 1833, 345)

12.   Richard Keynes, ed. *Charles Darwin's Zoology Notes & Specimens from the* Beagle (Cambridge: Cambridge University Press, 2000), 13.

13.   Himmelfarb, p. 71.

14.   The Concordance to *The Origin of Species* notes "wonder" six times, "wonderful" 28 times, "wonderfully" seven times, and "wondrous" once.

15.   Michael Ghiselin, *The Triumph of the Darwinian Method* (Chicago: University of Chicago Press, 1969).

16.   Max Weber, "Science as a Vocation," in *From Max Weber*, H. H. Gerth and C. Wright Mills (eds.) (Oxford: Oxford University Press, 1958, 1946), 129–58.

3

The Prose of *On the Origin of Species*

It is style that makes us believe in a thing—nothing but style.

Oscar Wilde

The art of the *Origin* is distinctly different from that of the *Journal of Researches*, and yet much of it depends on the powers we have seen Darwin developing in the earlier book: his extraordinary power to "see," to tell stories that turn facts into histories, doggedly to get all the particulars and to get them right, and his curiosity that turns all things into meanings not obviously visible on the surface. While, as "one long argument," the *Origin* would seem to depend entirely on the production of evidence, the logical tightness, *and* the careful thematic organization of the materials, it is equally dependent upon metaphors that demonstrate Darwin's remarkable capacity to recognize similarities across space and time, and upon thought experiments, which are really probabilistic stories; and while nothing of the larger argument would seem to require moments of Romantic intensity of the sort that are common in the *Journal of Researches*, passages of loving attention to the billowing of spiders' webs, or expressions of awe at the grandeur and complexity of natural phenomena, such elements become important parts of Darwin's tightly ordered and rhetorically shrewd presentation of his thesis. The *Origin* is at the same time encyclopedic and rigorously organized, comprehensive and selective, grotesque and rational, objective and personal, logically gymnastic and efficient, and deeply felt. In the *Origin* the powers of the scientist and the powers of the writer/artist converge.

The *Origin*'s success depends on Darwin's ability to convince readers of the superiority of its argument for descent by modification through natural selection to the theory of independent creation; and thus while it employs as evidence and as imaginative hypothesis a broad range of narratives, its organization is logical—and despite rumors to the contrary, tightly logical, if elaborated often with an abundance that might obscure its consecutiveness. The qualities of the grotesque and the aberrant, which are features of the world as Darwin describes it, in effect replace the qualities of order and implied intentional design that are the basis of natural theology's argument for creationism. In a remarkable inversion, the incongruous, the contradictory, the bizarre become a mark of scientific accuracy and essential to explain much of the world's apparent orderliness. This is the primary inversion that characterizes Darwin's great argument, but there are so many directions, and often hitherto unpredictable ones, in which Darwin needs to go that the rigor of his logic might be obscured.

To make his case absolutely, he must make it universal. And thus he searches his own mind and the literature for *any* possible exception or objection, and he addresses each one as fully as he can. Gillian Beer has argued that Darwin's theory is "essentially multivalent," and that "it renounces a Descartian clarity, or univocality" (p. 9). While this certainly, in the long run, has to be true because Darwin cannot stop with theory, but needs to make his case through multiple particulars, the argument has to have a rigorous, perhaps even "Cartesian" basis. Any phenomenon that doesn't fit would, as he says, be fatal to his theory. He had then to be both rational and encyclopedic; and Beer is certainly right that he could not keep the mass of possible facts and arguments completely under control.

Nevertheless, Darwin strains throughout the *Origin* for something like Cartesian clarity, and much of the literary tension in the book derives from the alternating commitment to achieve it and the recognition that the world just won't sit still for it, that nature is never that simple. "From so simple a beginning," he famously concludes the book, "endless forms most beautiful and most wonderful have been, and are being, evolved" (p. 490). The simplicity is attractive; the complications, the "endless forms,"—even, and often particularly, the strange ones—are "beautiful and wonderful."

Nevertheless, the first four chapters have about them an aura of precise and logically irrefutable argument. It is all there in those four chapters, and they are clear enough—metaphors included. Darwin wanted to make it appear a simple matter, and the first four chapters almost succeed in doing that; but most of the book is devoted to "Difficulties on Theory," problems he anticipated, where the complications underlying the simplicity are engaged. Beyond the "difficulties," however, with the particulars and against the sheer logic with which, repeatedly, Darwin overcomes them, he knew the larger difficulty of getting his readers to *feel* the reality of what had been rationally demonstrated. There were many objections, with which Darwin would grapple for the rest of his life, but the power of the *Origin* depends finally on its success in creating a way of seeing and feeling that allowed readers to experience the world freshly, to absorb a sense of dazzling multiplicity beyond the complete comprehension of any observer or participant, of sensing, even without fully understanding, the new reality. As for Gopnik, so Darwin hoped it would be for readers who would feel the world "vibrate" under the spell of his argument.

And it is here in particular that Darwin the writer, the artist, emerges. The *Origin* must not only make a powerful evidential and logical case for descent by modification through natural selection, it must make the readers experience the wonder of the world under the hand of natural selection. It is for this reason that, as David DePew puts it, "From start to finish the speaker interrupts his exposition to put himself in the reader's place . . . he seems to take his readers' worries on himself."[1] Darwin becomes a narrator of his scientific argument who knows what it *feels like* to encounter the sometimes overwhelming facts and ideas he describes, who has experienced what readers, confronted by these facts and ideas, will be thinking and feeling, and who seeks ways to lead them through their doubts and reluctance, as he had moved himself, to his own exhilarating sense of the world newly perceived.

Very early in the book, in the third chapter, on "The Struggle for Existence," he announces the problem and the strategy: "Nothing is easier than to admit in words the truth of the universal struggle for life, or more difficult—at least I have found it so—than constantly to bear this conclusion in mind" (p. 62). "At least I have found it so," says the already converted Darwin, inviting the reader to "bear this conclusion in mind," even against the difficulty— words against experience. It is the central rhetorical problem of the *Origin*. The words have to lead the reader to what till then would have seemed a totally counter-intuitve understanding of the natural world, and they will do so in part by having a "narrator," let us call him, who knows what it is that the skeptical reader will be feeling. Looking at the phenomena in an intuitive way, the reader will be reluctant to accept Darwin's interpretations. The experience of the *Origin* requires us to come to terms with the counterintuitive.[2] Darwin has only language with which to impress on his readers' minds the reality of what they cannot see. Language has to do the work of experience.

There is of course the scientist's rhetorical problem, which Darwin concedes: "in order to treat this subject [whether organic beings in a state of nature are subject to variation] at all properly, a long catalogue of dry facts should be given" (p. 44). He knows he has to fill the *Origin* with them, and he keeps letting the reader know that he has many more such "dry" facts that he would trot out if the *Origin* weren't an "abstract" and he had more time and space. Darwin projects his sympathetic awareness that they will seem "dry" to his readers, although we have seen how much pleasure they give him. "Dry," like many other undramatic

words in the book, implies Darwin's personal appearance in his
ostensibly objective text. It is not the scientist's word, but the word
the narrator finds to represent the reader's response.

His argument must make its way against everyone's intuitive
sense of what the world is like; anticipating and sharing the readers' feelings are part of the argument itself. Another clause, "At
least I have found it so," is another intrusion of the narrator, and in
implying that he is not speaking for everyone, he paradoxically is,
for he is identifying himself with the "everyone" out there who is
reacting normally. The paradoxical nature of these intrusions is
made overt with a remarkable comment later on. While sharing
the reader's way of thinking and feeling, he urges the reader to
make a kind of leap: "his reason ought to conquer his imagination"
(p. 188). Startlingly, Darwin once again enters the text personally,
implying his own struggle to get past the normal intuitive response
to things and push on to think them through. He associates an idea
that is counter-intuitive and thus instinctively registered as outside
the rational, just with the rational. There is a whole submerged
narrative here: if we only force ourselves to follow what the facts
and reasonable inference from them require, we will find ourselves
in a counter-intuitive world (which is nonetheless "rational").
Everything is upside down, though Darwin doesn't emphasize
the strangeness, but is in fact busy demystifying the strange by
identifying it as accessible through reason. The complexity of the
argument and the narrative is controlled and to a certain degree
disguised by a rhetoric that dramatizes Darwin's participation with
the reader in the effort: "I have felt the difficulty far too keenly to
be surprised at any degree of hesitation in extending the principle
of natural selection to such startling lengths" (p. 188). Reason is
"startling."

How can we use words to get beyond their limits? We can
"admit in words the truth," but can we genuinely know it, feel it
to the tips of our fingers—that, for example, the struggle for
existence is going on even as we watch a serene and placid nature,
a nature "bright with gladness," but in which "the birds which are

idly singing round us mostly live on insects or seeds, and are thus constantly destroying life"? Can we *see*, among the pleasures of an abundant spring time, that those very lovely birds and their "nestlings" in the garden, enlivening it with their song, "are destroyed by birds and beasts of prey," that "though food may be now superabundant, it is not so at all seasons of each recurring year"? His readers will have seen the birds, enjoyed their springtime singing and their colors and the serenity their presence induces; they will also, however, have known the bare ruined choirs of winter. Darwin puts the experiences together, attempting to force the reader to make connections, as he had long before learned to do in asking endless questions about every natural phenomenon. He reminds the reader of the two different conditions, and then connects them. The happy bird lives, probably unhappily, through periods of dearth and cold; the happy bird is subject to predation. Thus readers have to be reminded to *see* what is not immediately visible but what they in fact know, to consider the two different conditions, and then to draw necessary inferences from the connection. The counter-intuitive fact emerges from two common facts, both inescapable and disturbing to the lyric expectations the idly singing birds evoke.

Through all this, the "reasonable" argument remains clear. To get an adequate sense of the richness and difficulty of Darwin's art, it is important to follow those first chapters and recognize that argument, which in its reasonableness is to conquer the imagination and take us to a probable condition that seems at first sight improbable. The argument is built from many details that it would be distracting to attend to here (as, for example, in the first chapter on domestic selection, Darwin shows how there is always "unconscious" selection going on among breeders as they regularly breed only the "best" of each animal); but it is important to note that for the most part, Darwin's method is to replicate in his prose the processes that he sees in nature—long, gradual movements that in the end produce extraordinary and apparently "catastrophic" differences. He is always invoking facts (the word "fact" or "facts" in the concordance to the first edition occurs close to 300 times!), and often he either promises more facts, or swears that he has them but that this "abstract" is the wrong place to give them.[3] The prose works rather as he describes the workings of domestic selection: "its importance consists in the great effect produced by the accumulation in one direction, during successive generations, of differences absolutely inappreciable by an uneducated eye" (p. 32). So natural selection works, so Darwin's prose often works.

In the very first chapter, Darwin observes that under domestication, animals can be bred to almost any degree because every generation produces variations, and breeders select from nature's still unpredictable (but there must be a law we don't know yet, thinks Darwin) variants to produce the characteristics they prefer in domestic animals: "under domestication, it may be truly said that the whole organization becomes in some degree plastic" (p. 80). In the second chapter, Darwin asks whether variations can happen in nature as they do with domestic animals, and of course goes on to show that they can, so he needs to show that natural "species" manifest irregularities and variations with surprising frequency, to such a degree that naturalists often disagree on what constitutes a species, what a variety. So unstable are species that Darwin looks at them "as only strongly-marked and well-defined varieties"(p. 55). The formulation is startling, the evidential support for it powerful. Since then variation does occur in nature, the next question is how? In domestic selection we have the breeder selecting, but what selects in nature? Of course, the traditional explanation, insofar as that question was addressed, was the Intelligent Designer, but Darwin looks *inside* of nature for his answer, and in the third chapter he locates that selective force in "The Struggle of Existence"—the tendency of every species to grow "geometrically" (Malthus, of course, lies behind Darwin's language here), and the indisputable fact that a very large proportion of all living things born do not survive to breed. Everywhere in nature there is a superabundance of eggs, and seeds, and births, such that unchecked any organism, even the slow-breeding elephant, would completely overrun the world in a relatively few years. Since elephants don't do that, Darwin searches for the "check." He discovers that "the structure of every organic being is related, in the most essential yet often hidden manner, to that of all other organic beings, with which it comes into competition, for food or residence, or from which it has to escape, or on which it preys" (p. 77). Here is the theory that grows from and justifies Darwin's preoccupation with context right from the start of his career. So finally, in chapter four, Darwin shows how this combination of facts makes nature itself the selector out of all that abundance and excess, and a selector more effective than the human breeder in domestic production. This selector is "Natural Selection" (p. 81).

That's the argument. One might respond as Huxley did, "How extremely stupid not to have thought of that!"[4] Since it is built so strongly on indisputable facts it might seem just the work of reason, but it is the effect of great imaginative leaps, inferences from the facts, and a profound analogy (between variation under domestication and the condition of "species" in nature). Yet Darwin wanted more time to make the case for his startling conclusions until his hand was famously forced by Wallace. Still, he works as gradually as he can inside the frenetic haste with which he put together the *Origin* and he delays as long as possible actually providing his summary, which appears at last near the end of the fourth chapter, in a series of carefully modulated and rhythmically balanced "if" clauses, each one implicitly referring to arguments made in the previous chapters, and each one requiring a logical concession from the reader. If any sentence in the book exemplifies Darwin's gradualist approach and justifies his defensive assertion in his *Autobiography* that he must have "some powers of reasoning," it is this.

If during the long course of ages and under varying conditions of life, organic beings vary at all in the several parts of their organization, and I think this cannot be disputed; if there be, owing to the high geometrical powers of increase of each species, at some age, season, or year, a severe struggle for life, and this certainly cannot be disputed; then, considering the infinite complexity of the relations of all organic beings to each other and to their conditions of existence, causing an infinite diversity in structure, constitution, and habits, to be advantageous to them, I think it would be a most extraordinary fact if no variation ever had occurred useful to each being's own welfare, in the same way as so many variations have occurred useful to man. But if variations useful to any organic being do occur, assuredly individuals thus characterized will have the best chance of being preserved in the struggle for life; and from the strong principle of inheritance they will tend to produce offspring similarly characterized. This principle of preservation, I have called, for the sake of brevity, Natural Selection. (p. 127)

It is an extraordinarily sensible piece of argument, taking one by one the indisputable facts: "organic beings vary," there are "high geometrical powers of increase," at least "at some season" there is "a struggle for life." Who would be willing to argue this against so distinguished a scientist, who has invoked along the way many witnesses to these facts? The next move, though clearly at least partly empirical, moves to a slightly more extravagant and risk-taking level. The relations "of all organic beings to each other" cause an "infinite diversity." Certainly Darwin has demonstrated that the complications in relationships in nature might well be considered "infinite," although the word leaps into hyperbole. But by the time he gets through all his "ifs," and draws his conclusions, he has completely turned the tables. It is no longer Darwin having to prove the improbable, but the objectors to his idea having to prove "a most extraordinary fact." Probability has shifted. It is the traditional idea of the permanence and impermeability of species that is not rational. The reason has conquered the imagination. Darwin shifts the burden of proof.

Nobody can reasonably object to the facts. And Darwin seems always to be concessive. While he willingly concedes that there are lots of variations that are of no particular value, he quite reasonably and apparently modestly suggests that it would be really "extraordinary" if *no* useful variations ever emerged. And *if* any such do—as they just about must—then all the rest he wants to argue follows. By way of the rhythms of his logic he takes us to a condition where the combination of these indisputable facts *requires* his conclusion.

While the procedure of argument is beautifully clear, and well balanced, in a way that points back to the rhetorical modes dominant in English prose in the eighteenth century, it is important to remind ourselves—to make sure that we are not merely sentimentalizing and prettifying—that what lies behind the argument is a process that not only is invisible and counter-intuitive, but that depends on multiplicity, destruction, and death. Darwin did always aspire to Cartesian clarity, but is regularly confronted by a complexity that, he confesses, entails endless complications. So it is all the more crucial that he be able to fall back on what cannot be "disputed," even by the most religiously orthodox of scientists.

In the very tricky last sentence of this quotation, the tables now turned, Darwin can talk of "Natural Selection," not as a force of death, but as a "principle of preservation." Natural Selection is not defined so as to emphasize its method of "choosing," that is, by killing every non-adaptive organism (although we know that's what it does). Instead, it is defined as a very positive "principle," one that preserves rather than destroys. Note how different this introduction of the idea is from the way it has been spun by most readers and by science itself. It is not Herbert Spencer's phrase, "Survival of the Fittest"—which Darwin, however reluctantly, agreed to let stand with "Natural Selection" from the fifth edition on. "Survival" implies at best a negative condition, plus struggle and competition; "preservation" implies a positive function, and mutes the violence that "survival" rather aggressively implies. The great and dominant metaphor of the book, "Natural Selection," is more maternal than military.

This point is a very familiar one, but attending to it once more is one of the surest ways to recognize how intensely the *Origin* is written, and how Darwin is willing at any given moment to burst

into metaphor that resonates well beyond literal meaning. Robert
Richards has been particularly astute in demonstrating that Darwin's presentation of natural selection is *not* mechanical, and *not*
negative.[5] Let's look for a moment at Darwin's summary of the
workings of natural selection after he first introduces that generous
Lady in chapter 4 of the *Origin*:

It may be said that natural selection is daily and hourly scrutinizing, through
the world, every variation, even the slightest; rejecting that which is bad,
preserving and adding up all that is good; silently and insensibly working,
whenever and wherever opportunity offers, at the improvement of each
organic being in relation to its organic and inorganic conditions of life. (p. 84)

Working "silently and insensibly" natural selection sibilantly tends
to her subjects. With rhythmic balance she rejects what is bad,
adds up "all that is good." And she works with the care of a deeply
responsible mother, allowing nothing of life to evade her scrutiny,
and watching always and everywhere the myriad relations by
which all living things exist. "She," as she is regularly called in
the passages leading to this summary, is characterized as a creature
more concerned for others than for herself, for while "man selects
only for his own good," she selects "only for that being which she
tends" (p. 83). She is a wonderful mother. If we add up her virtues
we find that she places all beings where the conditions of life are
best suited; she makes distinctions that humans do not make and
provides for each being the right food and the right exercise.
"Under nature, the slightest difference of structure or constitution
may well turn the nicely-balanced scale in the struggle for life, and
so be preserved." I confess to having cheated here by leaving out of
these citations the brief passages that imply that natural selection
will "rigidly destroy all inferior animals," even as she protects them
during each varying season. They are reminders that natural
selection's is a very tough love, that its workings mean death as
much as they mean life. But in Darwin's conception of her, those

deaths are in the interests of the creatures for which she cares, that is, they are in the interests of life.

Having made natural selection a loving and highly moral person, Darwin bursts into almost biblical lyricism when he exclaims, "How fleeting are the wishes and efforts of man! how short his time!" The whole sequence has the texture not of scientific tract but of literature charged with meanings beyond the literal, and registering nuanced feeling even as it attempts strenuously to make a careful argument.

The frame in which these meanings and feelings are contained should by now be familiar from Darwin's work on the *Beagle*, for Darwin's first scientific objective is to make what seems "wonderful" comprehensible in the terms of common life and "reason," to banish the wonder with which he begins and which proves so often to be the impetus of his argument. "Can we wonder, then," he asks, "that nature's productions should be far 'truer' in character than man's productions; that they should be infinitely better adapted to the most complex conditions of life, and should plainly bear the stamp of far higher workmanship?" (p. 84). Fending off the wonder, he invokes it. Showing that nature is much more skilled than man, he produces a wonderful fact to ward off an initial blind wonder.

It is obviously not accidental that Darwin talks of workmanship here;[6] it is part of his early conception of nature, and one is reminded of his saying in his *Autobiography* that William Paley's *Natural Theology* was one of the most important books in the shaping of his mind. But in Darwin's work the design comes from the nature that is vastly superior to man in intelligence, awesome but explicable. Darwin's nature, rather than being emptied out of meaning because Darwin does not invoke God, is filled with meaning that also transcends human powers, but does not take human powers as a model for the intelligence it finds in nature. It is nature that does the work of Paley's God.

No wonder that Gillian Beer can so effectively argue that Darwin's work is touched by a "remnant of the mythical" (Beer, 3). Natural selection emerges as a Godlike figure, caring for every hair on the head and all the slightest details of life that humans do not and often cannot notice. Darwin's theory, proposing to resolve the "mystery of mysteries"—that is, the origin of species—implicitly addresses the origins of man (although as we know Darwin was careful to skirt direct statement about our species until *The Descent of Man*). Robert J. Richards describes how Darwin's "conception of natural selection sprung from the head of a divinized nature" (*Romantic Conception*, 5). For Darwin, nature and natural selection, at least in his early conception of them, are neither machine-like, nor do they depend for their working on some previous intelligent designer. It is only that nature, as he describes her work, *seems* to be intelligent.

We can say that we know what Darwin really meant. He claims in later revisions that "everyone knows what is meant." The metaphor is only a vehicle that Darwin famously tried to explain away.[7] But it matters greatly that "natural selection" has survived as a living metaphor, and that Darwin regularly personified it in ways that make it intellectually and morally superior to humans, certainly not frightening or demoralizing. It emerges not as a mere process, but as a living entity, who is more careful and attentive than man—"Man can act only on external and visible characters: nature cares nothing for appearances, except in so far as they may be useful to any being. She can act on every internal organ, or every shade of constitutional difference, on the whole machinery of life." This agent, reduced by Daniel Dennett to an "algorithm,"[8] and by all of evolutionary biology to a mere indifferent process, is in Darwin's prose a woman, perhaps a goddess, who "acts," and one who is morally and intellectually superior to mankind because she is not distracted by mere appearances. The language Darwin finds to describe "her" is applicable to moral beings.

Consider again the moral implications of "Man selects only for his own good; Nature only for that of the being which she tends" (p. 83). Natural selection is selfless; man is selfish. If the *Origin* had been a novel, "Natural Selection" would be the good woman, who is always helping others and usually gets to marry the hero. "Man," the self absorbed villain, exploits others for his own interests.

Darwin's representation of her makes natural selection an intellectual and moral agent, rather than a brute mechanical process. The
rhetorical force of this imaginative creation is particularly strong
because it is obviously not a metaphor laid over an original
abstract conception, but intrinsic to the conception itself.[9] When
Darwin tried to explain his idea to Asa Gray in 1857 (in a letter that
was ultimately used in his pairing with Wallace's essay at the
Linnaean society meeting in 1858), he puts it this way, with Natural
Selection designated in the masculine: "Now suppose there was a
being, who did not judge by mere external appearance, but could
study the whole internal organization—who never was capricious—who should go on selecting for one end during millions
of generations, who will say what he might not effect!" (*CCD*, 5
September 1857, vol. 6, p. 447). In emphasizing this quasi-deifica-
tion of the process, I am not denying the use of the idea in modern
evolutionary biology, but only insisting, with Beer and Richards,
that coming to terms with what Darwin really said entails coming
to terms with the *way* he developed his ideas and the language he
found to express them. The living, choosing, non-capricious being
lurks in the pages of the *Origin* down through the sixth edition,
even when Darwin had already agreed to incorporate Spencer's
phrase, "survival of the fittest."

At the start of her crucial study, Beer claims that "Ideas pass
more rapidly into the state of assumptions when they are *unread*.
Reading is an essentially question-raising procedure" (p. 6). The
ideas can indeed be lifted out of the texts, as Darwin's have so
usefully been, particularly by scientists, and sometimes not so
usefully by non-scientists. But those ideas are only a part of the
question here: the mind-changing "experience" is crucial and
makes the ideas possible. The Romantic Darwin, who exclaims
about natural selection, "who will say what he might not effect!"
thrills with excitement of his discovery of this alternate god.

The simplicity of Darwin's basic argument, then, depends on an
almost endlessly spiraling (Darwin was fond of the hyperbole,
"infinite," which turns up around thirty times in the first edition)

complexity of relationship and extension of time. The mythic elements of his origin saga require in place of God some "infinitude," and the Romantic elements of Darwin's writing have much to do with his persistent, and inevitably incomplete, quest to come to terms with that infinitude. As its metaphors reach beyond the limits of the visible, beyond, Darwin might say, the reach of imagination itself, his arguments rely on sometimes elaborate metaphors, gestures of *not* knowing, exclamations of wonder, that is, on elements that might well be associated with poetry or fiction.

The "structure of every organic being," he says, "is related, in the most essential yet often hidden manner, to that of all other organic beings" (p. 77). The hidden essence to be detected is what Darwin tries to display in his writing, and what drives him to long sequences of speculative narrative. The narratives provide possible, even probable, explanations, but the unknowns remain, and in being driven to write those narratives, Darwin consistently gives the impression that man is inadequate to this richness, not least because there are so many elements of the theory that he cannot reduce to irrefutable, empirically grounded law, though he would like to. A large part of the effect of the *Origin* depends on the quite remarkable number of "unknowns" that at once baffle and do service.

A litany of "unknowns" might suggest how on the one hand Darwin must leave himself vulnerable to critique, and on the other, his brilliance and ingenuity allow him to use the unknown as a supporting part of his argument. A trip through the *Concordance* to the *Origin* is revelatory.[10] I count 53 "unknowns"—among them, "the quite unknown or dimly seen laws of variation," whose result is "infinitely complex" (p. 12); and the "unknown laws governing inheritance" (p. 13); or the "modifications of structure," which "may be wholly due to unknown laws of correlated growth" (p. 146); or "the complex laws governing the facility of first crosses," which "are incidental on unknown differences" (p. 263). There are unknown common parents, unknown periods of time, unknown deeds of violence, and unknown means of dispersal.

Behind the unknown lies always the "infinite," which inheres also in all the individual details, in a way that Darwin tries to reproduce in his writing. His sense of the infinite complexity of relationships is anchored to particular organisms, and those organisms, as in a sacred world but here in a world governed rather by natural selection, are full of significance and of beauty.

> We see these beautiful co-adaptations most plainly in the woodpecker and missletoe; and only a little less plainly in the humblest parasite which clings to the hairs of a quadruped or feathers of a bird; in the structure of the beetle which dives through the water; in the plumed seed which is wafted by the gentlest breeze; in short, we see beautiful adaptations everywhere and in every part of the organic world. (60-1)

Darwin's readers are to register this extraordinary richness and beauty at the same time that they must understand that the beauty and the meaning are not designed for them but for each of the organisms whose wonders we are all privileged to learn about. Darwin's eyes take us beyond the easily visible, with his litany of examples equating birds with the parasites on their feathers, with beetles, and seeds. It is notorious that there is a persistent force deflationary of human pride in Darwin's arguments, but that force has a revelatory epistemological and moral implication. Darwin forces us to see what is not obviously there to be seen, and to value everything that we see. There is beauty in "every part of the organic world."

While that deflationary implication of his theory has offended generations of believers, it emerges from Darwin's pen as a profoundly ethical development. Learning the importance and reality of others, not assuming that the world is made for us, understanding that every organism (with implications for our human organisms always unspoken but latent in the *Origin*) is connected to other organisms, visible and invisible, all around us, changes our relation to the world and challenges the implicit human exceptionalism of our relation to the earth and its inhabitants. Darwin's antipathy to the anthropocentric arguments of Natural Theology is well known—if it were true that any single organism or part of an organism were there not for the organism's sake but for the sake of mankind, his theory, he believed, would be destroyed. "Some naturalists," he says, have protested "against the utililitarian doctrine that every detail of structure has been produced for the good of its possessor" and believe "that very many structures have been created for beauty in the eyes of man, or for mere variety. This doctrine, if true, would be absolutely fatal to my theory" (p. 199). Which is to say that to believe in natural selection requires a strong moral turn from self-centeredness.

At one point, in his famous comparison of the eye to a telescope, which comments ironically on Paley's use of the same analogy, Darwin pauses to ask, "have we any right to assume that the Creator works by intellectual powers like those of man?" (p. 188). Here again, we can recognize the moral energy that helps drive Darwin's formulation of his theory, which is implicitly, in virtually all of its details, an attack on vanity. With a humility that matches his own instinctive modesty, Darwin insists on the limits of knowledge and thereby turns the moral tables. As he had made the argument against the changing nature of species seem improbable, because the failure of things to change, given the facts he had produced, would have been "extraordinary," so here he turns the conventional theological explanation into mere prideful foolishness. The real moral stance is one that recognizes humbly the impossibility for humans to understand the ways of God. Darwin holds God to a higher standard than do conventional believers, and without claiming it, he depends on the moral superiority of the naturalist's stance. Recognize the limits of our knowledge, for that affirms both the infinite multiplicity of the world and the humility of humans aspiring to understand it.

But Darwin's deflation of human-centered views of nature goes well beyond the critique of natural theology, for it entails a persistent invidious comparison between man's powers and nature's (just as religions traditionally make that comparison between

man's powers and God's). After having demonstrated how effective selection of domestic stock by farmers and breeders can be in
producing astonishingly different races from the same parent
stock, he concludes: "If feeble man can do much by his powers
of artificial selection, I can see no limit to the amount of change, to
the beauty and infinite complexity of the coadaptations between all
organic beings, one with another and with their physical conditions of life, which may be effected in the long course of time by
nature's power of selection" (p. 109). Nature, the dominant figure in
the *Origin*, is regularly shown to operate without *particular* regard
for the human, and with powers that transcend human intelligence:
"Natural Selection . . . is as immeasurably superior to man's feeble
efforts, as the works of Nature are to those of Art" (p. 61).

The complexity of nature always outgoes Darwin's strenuous
efforts to represent it. Both the art and the science of the book are,
however, supported by the complexity and by the limits of human
knowledge about nature, for those limits open possibilities of
hypothetical and probabilistic narratives that are themselves
quite convincing, often wonderful stories; and, on Darwin's legitimate understanding, they reasonably represent possibilities of
causes now in operation—Lyell's actualism. "Infinite" is perhaps
not merely a hyperbole but a considered estimation of the unlim-
ited ways in which organisms can relate to each other, to themselves, and to their environment. At the very start of chapter 4,
"Natural Selection," Darwin enjoins the reader: "Let it be borne in
mind how infinitely complex and close-fitting are the relations of
all organic beings to each other and to their physical conditions
of life" (p. 80). But obviously too, there is no way in nature to prove
the "infinite"; it is just that everywhere Darwin looks in his investigations of the history and relationships of the organisms he studies, he
finds connections that can't be resolved in a definitive way.

Although natural selection becomes "infinite," or at least infinitely complex, it is also the key to each organism's identity; while
it works on whole species, it works always directly on particular
individuals. One of the central forces in producing Darwin's

rhetoric and his art is evident in the assertion of "the infinite complexity of the relations of all organic beings to each other and to their conditions of existence," an assertion that appears, as we have already seen, in slightly different form throughout the *Origin.* Infinite variety, infinite diversity, infinite time—all of these make natural selection "infinitely" rich in producing its "infinitely complex relations to other organic beings and to external nature," all of which "will tend to the preservation of [the] individual" (p. 61).

That infinitude gives to natural selection much of its mythic force, and to the *Origin's* prose, as it emerges from empirical fact, its persistently hypothetical nature. While Darwin cannot possibly prove his case absolutely, no matter how plausible and even logical his explanatory, probable, narratives, the felt experience of the infinitude that the prose evokes compensates for the absence of the smoking gun. One might feel that infinitude as Darwin pushes the boundaries of time and organic abundance and interrela-tions—from parasites on bird feathers, to cirripedes and mollusks, to ants' nests and bee's hives and eagle's eyes and bears, to virtually every part of the organic kingdom. It is like sitting in on the new creation. These infinite extensions of life, which moves slowly but is everywhere, through fields of the known and the unknown, lead readers by way of incremental experience on the crucial steps toward acceptance of what Darwin offers, in his first four chap-ters—a simple, rational and empirical argument.

Well, it is never really that simple, and it is never merely rational. Here is how Darwin confronts the problem of what it is that leads particular organisms to extinction. The fact of extinction is itself a stunning one, especially given a creationist reading that each species has been individually created by God, and as Darwin approaches the problem he confesses its difficulty: "No one I think can have marveled more at the extinction of species than I have done" (p. 318). The move, beginning with wonder, is by now familiar, and it is repeated in many particulars. He was, for example, "filled with astonishment" when he found a horse's tooth "embed-ded with the remains of Mastadon, Megatherium, Toxodon, and other extinct monsters" (p. 318). How could it be that an older species of horse, to which the tooth belonged, went extinct in an environment in which the modern horse thrived and multiplied? "I asked myself, what could so recently have exterminated the former horse under conditions of life apparently so favourable?" What must follow, on the Darwinian pattern is the abolition of "marvel" and "astonishment" by finding recognizable causes— what Lyell and other scientists of his time would have called "*verae causae.*"

He does it by telling a story—an act of imagination that we would now call more tamely a thought experiment—built out of a combination of actualism, ignorance, analogy, and strikingly subtle reasoning. Positing the existence of a very rare horse living today, he considers how we might explain its rarity: "If we ask ourselves why this or that species is rare, we answer that something is unfavourable in its conditions of life; but what that something is we can hardly ever tell." On the analogy, then, with living animals like the domestic horse, we know that the fossil horse, had it been living in favourable conditions, "would in a very few years have stocked the whole continent." But even if we were not able to perceive the development of unfavorable conditions they would, "however slowly," have made the horse "rarer and rarer, and finally extinct" (p. 319). If that would be the case now, why should there be astonishment at finding an extinct species of horse? It all sounds wonderfully obvious and simple. And so he exclaims, "how utterly groundless was my astonishment" (p. 319).

Such a story, typically for Darwin, builds on his awareness of what is not visible, of the fact that, as we watch the birds contentedly singing, the trees are not always full, and birds of prey are always ready to strike. "It is most difficult to remember that the increase of every living being is constantly being checked by *unperceived* injurious agencies" (my emphasis). Once again, Darwin's world is filled with the unknown and the unperceived, which, as an imaginative writer, he must detect. And as an imaginative writer he works constantly to awaken the reader to the reality of what they cannot see, to make them feel meanings that mere passive observation cannot suggest. The stories Darwin tells are of course works of the imagination, but of a specially tutored imagination that builds its fictions out of the most likely realities. All his possible explanations are this-worldly, recognizable to common sense after all, but nonetheless wonderful for that. Extinction is undoubtedly the result of a great entanglement of unfavorable circumstances, perhaps each one tiny, but together, over the long passage of time, deadly. And thus, Darwin concludes this discussion with what is almost a refrain of his prose: "We need not marvel at extinction":

If we must marvel, let it be at our presumption in imagining for a moment that we understand the many complex contingencies, on which the existence of each species depends. If we forget for an instant that each species tends to increase inordinately, and that some check is always in action, yet seldom perceived by us, the whole economy of nature will be utterly obscured. Whenever we can precisely say why this species is more abundant in individuals than that; why this species and not another can be naturalised in a given country; then, and not till then, we may justly feel surprise why we cannot account for the extinction of this particular species or group of species. (p. 322)

There are this-worldly explanations of everything, and only when such explanations are exhausted are we entitled to the wonder with which we began. Note how this summary passage employs so forcefully the trope of not knowing as a means to affirmation. Note that it takes the "marveling" at the fact of extinction as something to be dismissed, but at the same moment invokes the experience of marveling for another even moralizing purpose—let us marvel at our arrogance and presumption in thinking that we can understand the "complex contingencies" of relationship that Darwin has been emphasizing throughout. We begin with the inevitable incapacity to comprehend entirely the infinite and indeed marvelous intricacies and entanglements of relationship.

In the face of this complexity, Darwin in effect asks the reader simply to accede to the fact of extinction, and to the fact that we cannot account for the extinction of any "particular species or group of species."

But he has already supplied us with many possible explanations, if with no specific single explanation of the extinction of any given species. In the intervals of the fossil record "there may have been much slow extermination," he says, speaking of the trilobites. There is some evidence—implicitly—of "sudden immigration" of new groups, and of a "correspondingly rapid" extermination of native groups, all sharing "some inferiority in common." There is no proven history, but some likely histories. So, Darwin goes on to condemn implicitly the arrogance of believing that such complex stories can be fully known for any species, while at the same time showing that there are in fact quite plausible stories that *might* be written, and he can make his point: "Thus, as it seems to me, the manner in which single species and whole groups of species become extinct, accords well with the theory of natural selection" (p. 322). We might want to call this sleight of hand, or of mind, but Darwin has really made a powerful case out of possible narratives that invoke overwhelming complexity—and out of implicit moral reproof. Marvel gives way to plausible explanation, but the plausible explanation invokes marvelous complexity.

There are many other examples of Darwin's endless curiosity and argumentative inventiveness that emerge in storytelling, but one is particularly striking in its range and proportions. He attempts to explain the "striking cases known of the same species living at distant points, without the apparent possibility of their having migrated from one to the other. It is indeed a remarkable fact. . . . " (p. 365). While the pattern of argument is the same, the imaginative reach is—may I say it Darwinianly—"astonishing." He notes that until at least 1747, the inference was that "the same species must have been independently created at several distinct points." Of course, this will not do for Darwin, for it doesn't answer the question but only avoids it. The assumption that this phenomenon must be explicable within the range of normal causation leads him on an epic imaginative journey built on what science really does know. The starting point is the Ice Age and the radical changes in climate of which the rocks of Europe, for example, give striking evidence: "The ruins of a house burnt by fire do not tell their tale more plainly, than do the mountains of Scotland and Wales, with their scored flanks, polished surfaces, and perched boulders, of the icy streams with which their valleys were lately filled" (p. 366). Darwin reads the traces, the rastro, as though the movements that produced them were still present to his eyes. He *sees* those icy streams just as he *saw* the violence that had produced the streams of stones.

At this point, Darwin moves to yet another thought experiment: "we shall follow the changes more readily, by supposing a new glacial period to come slowly on, and then pass away, as formerly occurred." He casually invents a new Ice Age, but with it Darwin can trace "how each more southern zone became fitted for arctic beings and ill-fitted for their former more temperate inhabitants." The effect would be the displacement of the old inhabitants by newer forms better adapted to the cold, some going extinct, some successfully moving south. At the same time, the inhabitants of cold mountain tops could descend to the equally frozen plains, and "by that time... we should have a uniform arctic fauna and flora, covering the central parts of Europe, as far south as the Alps and Pyrenees." The same effects would be produced in the New World. So far, we have an entirely made up but entirely plausible guess: if there had been such an Ice Age, this would have had to have been the effect.

And then, the next part of the story: as this "new glacial period" thawed "the arctic forms would retreat northward" and climb those mountains that retained the arctic conditions on which they thrive. And "Hence, when the warmth had fully returned, the same arctic species, which had lately lived in a body together on the lowlands of the Old and New Worlds, would be left isolated on distant mountain-summits (having been exterminated on all lesser heights) and in the arctic regions of both hemispheres" (p. 367). The whole narration is conducted in the quietest and most matter of fact way—first this, then that, then that. It all makes sense, but we have been reading a breathtakingly ambitious narrative. As I have suggested earlier, Darwin is not given to frequent lyric and Romantic outbursts, particularly not after his *Journal of Researches*, but he achieves sublime Romantic visions in a coolly rational tone of voice, and with the confidence of someone profoundly familiar with the way weather, geology, and animal behavior would go under circumstances of the kind he imagined. It is a sweeping story covering millions of years and vast areas of the earth. It is modestly affirmed, a reasonable guess, but in a fully naturalistic way it provides a plausible explanation of what had seemed unequivocally clear evidence of separate creation. There are of course more details and more complications, but the story, built on reading traces, seeing what is not visible, recognizing by analogy the way organisms connect with their environment, imagined consecutively, is of more than mythic proportions. The grandeur is there, after all.

The metaphorical nature of Darwin's overall enterprise is somehow contained, at least most of the time, in this modest and apparently objective prose.[11] But that Darwin thought metaphorically is clear, most obviously, in the diagram he produced in his "B" notebook in 1837 as he had begun his first questioning about evolution (see Figure 3.1). The most striking thing about the diagram is that it follows immediately Darwin's words, "I think"! Although on the diagram, Darwin tries to explain its implications in his characteristic notebook prose, it is clear that he "thought" evolution by seeing it; it came to him visually, in his own peculiar form of the famous "tree of life." The image spurs the language as metaphors spur meaning, but the image keeps unfolding possibilities (like the single diagram in the *Origin*, to be discussed in this chapter).

Here is one of the most famous of Darwin's visual metaphors: "The face of Nature may be compared to a yielding surface, with

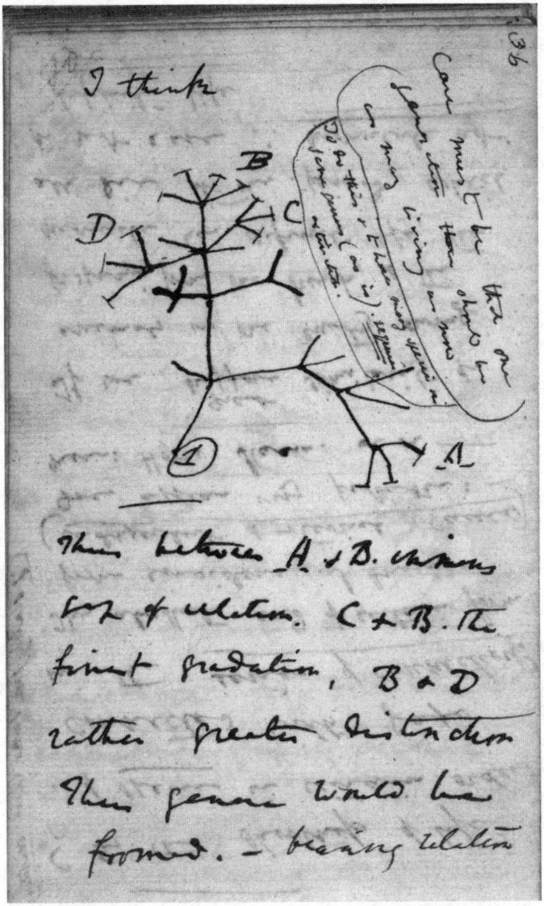

Figure 1. Darwin's first tree diagram from the "B" notebook in 1837, MS.DAR.121, p. 36. Reproduced by kind permission of the Syndics of Cambridge University Library. Beneath the diagram, Darwin writes, "Thus between A. and B. immense gap of relation. C & B. the finest gradation, B & D rather greater distinction. Thus genera would be formed.—bearing relation." (*Notebooks*, p. 180)

ten thousand sharp wedges packed close together and driven inwards by incessant blows, sometimes one wedge being struck, and then another with greater force" (p. 67). Just after reading Malthus, in the notebook entry contemporary with the famous treelike diagram with which Darwin first imagined natural selection, Darwin had used that metaphor earlier:

One may say there is a force like a hundred thousand wedges trying force <into> every kind of adapted structure into the gaps <of> the oeconomy of Nature, or rather forming gaps by thrusting out weaker ones. <<the final cause of all this wedgings, must be to sort out proper structure & adapt it to change...[12]

As Darwin found ways to articulate his theory, in the 1842 and 1844 drafts, the metaphor remained.[13] But by the sixth edition of the *Origin*, the metaphor was gone (Darwin having stricken through the passage in the first edition in his copy of the book). One might speculate on why it disappeared. Perhaps, in consonance with his decision to explain his metaphor of natural selection in unmetaphorical language, Darwin was worried by criticisms of his metaphors and this one was particularly striking. Or perhaps because he recognized that the metaphor too forcefully emphasized the violence and competition that his theory implied rather than other elements that might have suggested "mutual aid." But certainly, the metaphor, springing so immediately to his pen as he first formulated the idea of natural selection, reflects his view that the process he was describing was indeed violent. And it is worth noting that in each of the drafts (and in the first edition), the metaphor is preceded rather closely by Darwin's allusion to de Candolle: "De Candolle's war of nature," is Darwin's note in the 1842 draft (p. 36); "De Candolle, in an eloquent passage, has declared that all nature is at war" (p. 88). In the *Origin*, however, that allusion is rather softened: "The elder De Candolle and Lyell have largely and philosophically shown that all organic beings are exposed to severe competition" (p. 62).

The history of these various important metaphors in Darwin's development makes clear that Darwin knew what he was doing with them. In the first edition, he recognized that even his phrase, "struggle for existence," which he had found earlier in Lyell,[14] was metaphorical and as such opened out a wide range of possibilities:

I should premise that I use the term Struggle for Existence in a large and metaphorical sense, including dependence of one being on another, and including (which is more important) not only the life of the individual, but success in leaving progeny. Two canine animals in a time of dearth, may be truly said to struggle with each other which shall get food and live. But a plant on the edge of a desert is said to struggle for life against the drought, though more properly it should be said to be dependent on the moisture. A plant which annually produces a thousand seeds, of which on an average only one comes to maturity, may be more truly said to struggle with the plants of the same and other kinds which already clothe the ground. The missletoe is dependent on the apple and a few other trees, but can only in a far-fetched sense be said to struggle with these trees, for if too many of these parasites grow on the same tree it will languish and die. But several seedling missletoes, growing close together on the same branch, may more truly be said to struggle with each other. As the missletoe is disseminated by birds, its existence depends on birds; and it may metaphorically be said to struggle with other fruit-bearing plants, in order to tempt birds to devour and thus disseminate its seeds rather than those of other plants. In these several senses, which pass into each other, I use for convenience sake the general term of struggle for existence (pp. 62–3).

We have come here a long way from wedges. Richards has subtly interpreted this sequence and its implications, indicating how this broader understanding of struggle mitigates the harshness of "Malthusian pitilessness."[15] Struggle here encompasses interdependence, and acts of love—or at least procreation. It is another and earlier version of Darwin's persistent motif—the infinitely complex sets of relations that determine the life of any organism. Exploring the metaphorical nature of his own thinking, Darwin expands away from some metaphors, and turns to others, more complicated and subtle, to do their work.

It is no wonder, then, that Peter Kropotkin, in *Mutual Aid* (1902), begins his discussion under the topic, the "struggle for existence," and goes on to develop just those qualities in Darwin's more general metaphorical sense that most of Darwin's followers and opponents do not emphasize. He points to Darwin's argument in *The Descent of Man* about the importance of community, though he laments that Darwin himself tended on the whole, elsewhere, to treat "struggle for existence" in what he called the "narrow sense," that is, the sense of struggle as combat of individuals.[16] Nevertheless, Kropotkin is correct that there are other senses in Darwin's expanded, and non-wedgelike metaphor. Kropotkin seems also correct in saying that Darwin tended to develop the "narrow sense" more than the one that Kropotkin wished to emphasize, mutual aid, as for example, in the relation of bee to flower. For Kropotkin, while it is true that the quest for food produced competition and struggle, the "struggle" for progeny, which includes, of course, love, but also the mutual support of community, produced mutual aid. Neither Rousseau, for whom, Kropotkin says, nature was all peace and calm, nor Malthus, for whom it was dearth and struggle, were correct. But mutual aid is as much a condition for survival as competition for food. That sense of mutual aid is already there by the time of the writing of the *Origin*, although Darwin never ceases also to recognize the war of nature.

However we interpret it, it is evident that metaphor was central to the way Darwin thought about nature and the way he wrote. His ideas were embodied, and his prose is offered to his readers as both argument and experience. Finding metaphors that would do justice to his sense of multiplicity, diversity, and infinite relations among organisms was difficult, but it certainly makes sense that the notion of "entanglement" played so large a role in the final metaphor of the book—"It is interesting to contemplate a tangled bank." But because that beautiful final paragraph is so often invoked and discussed, I will leave that until my final chapter, except to note that its work is just to convey in a local and recognizable scene the enormous complexity and richness and, yes, "grandeur" of Darwin's nature—and of his prose.

Here I would like, instead, to pause over another crucial and familiar metaphor (or in this case, more properly, simile), to convey a sense of the affective power and, at the same time, scientific precision of his metaphorical imagination. Once again it is a problem of bringing together a sense of enormous complexity with a fundamentally simple intellectual argument. The tree of life, which we have seen was implicit in his first imagination of evolution, is a figure familiar throughout almost all of literary history, so there is nothing particularly striking about the fact that Darwin, in a book that has mythic implications, uses it as a means to explain as clearly as he can the significance of his theory. But the vehicle of this tree of life has the unusual virtue of a botanist's literal precision that gives peculiar strength to its tenor. Concluding his chapter on "Natural Selection," he produces the "simile," as he rightly calls it:

The affinities of all beings of the same class have sometimes been represented by a great tree. I believe this simile speaks the truth. The green and budding twigs may represent existing species; and those produced during each former year may represent the long succession of extinct species. At each period of growth all the growing twigs have tried to branch out on all sides, and to overtop and kill the surrounding twigs and branches, in the same manner as species and groups of species have tried to overmaster other species in the great battle for life. The limbs divided into great branches, and these into lesser and lesser branches, were themselves once, when the tree was small, budding twigs; and this connexion of the former and present buds by ramifying branches may well represent the classification of all extinct and living species in groups subordinate to groups. Of the many twigs which flourished when the tree was a mere bush, only two or three, now grown into great branches, yet survive and bear all the other branches; so with the species which lived during long-past geological periods, very few now have living and modified descendants. From the first growth of the tree, many a limb and branch has decayed and dropped off; and these lost branches of various sizes may represent those whole orders, families, and genera which have now no living representatives, and which are known to us only from having been found in a fossil state. As we here and there see a thin straggling branch springing from a fork low down in a tree, and which by some chance has been favoured and is still alive on its summit, so we occasionally see an animal like the Ornithrhyncus or Lepidosiren, which in some small degree connects by its affinities two large branches of life, and which has apparently been saved from fatal competition by having inhabited a protected station. As buds give rise by growth to fresh buds, and these, if vigorous, branch out and overtop on all sides many a feebler branch, so by generation I believe it has been with the great Tree of Life, which fills with its dead and broken branches the crust of the earth, and covers the surface with its ever branching and beautiful ramifications. (130–1)

This is not only precise, but it soars into something like lyrical beauty. Here the word "ramifications," etymologically accurate in its root meaning of "branching," if not obviously beautiful, sustains the parallel between literal branching of a literal tree and the counter-intuitive branching of varieties and species. Here, too, Darwin seems to have attended to things like alliteration and internal rhyme, in the "crust" covering the "surface," in the "broken branches," issuing in "beautiful ramifications." But whether one attends to such poetic nuance or not, it is clear in this passage that Darwin's vision is fundamentally metaphorical, capable of finding in the precisely observed particulars of the natural world graphic and verbally sensitive representations of his encompassing general vision. The simile even seems to suggest that all natural forms, in their various ways, repeat the curves and turns of natural selection, rather like what we now call fractals. The pattern of natural selection is played out in the growth of trees, as, implicitly, it seems to be played out in the very bodies of all organisms, each one spreading and branching, leaving as mere vestiges formerly important parts.

Although, as Jonathan Smith[17] points out, the *Origin* is strikingly without illustrations, something rather unusual for a book on natural history, there is a distinct visual image here: the famous diagram in the fourth chapter of the *Origin* offers several "trees" rather like the more literal one described in this passage. Darwin's initial insight by way of diagram (see Figure 3.1.) laid the ground for the figure in the *Origin* and suggests that Darwin truly "saw" this metaphorical branching. Figure 3.2, of course, represents nothing materially observable in nature; it is not part of the empirical evidence for the theory. Rather, it represents a "thought experiment" that relies on the metaphor of the tree of life with its branching, and allows Darwin to play out with multiple variations in several chapters the various way in which natural selection proceeds on every time scale.

This unprepossessing image opens out into many pages of possible interpretations, into many "stories." Darwin returns to discuss the diagram in later chapters as well. The visual image of the tree, now transformed into something that seems a bit more orderly, translates into an extraordinary number of possible interpretations, all, of course, consistent with each other, and yet all different narratives and different ways to explain the complexity of the workings of natural selection, or "divergence of character," and of speciation.

Finally, here, I want to look at a remarkable sequence in the chapter on "The Struggle for Existence" that might serve as a fair representative of Darwin's qualities as a writer. It manifests the

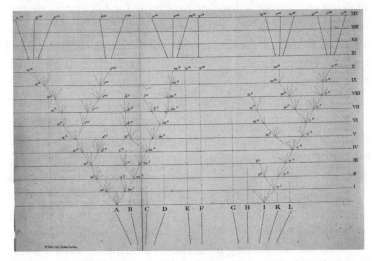

Figure 2. Darwin's diagram, inserted into chapter IV of *On the Origin of Species* (London: John Murray, 1859), Syn.7.85.6. Reproduced by kind permission of the Syndics of Cambridge University Library.

complexity and subtlety of his thought and the way in which his profound commitment to the most accurate and direct explanation of his ideas and description of the world reflects his rare combination of enormous intellectual ambition and domestic modesty, of analogical imagination, and the capacity to detect traces and connections everywhere, and of lyrical engagement with the enormous complexities and wonders of the nature he devoted his life to trying to understand. It is one of those passages that, while sustaining the wonder-explanation, wonder-pattern of so much of the *Journal of Researches* and the *Origin*, emphasizes, with vital particularity, the infinite interdependence of organisms on each other, on the weather, on time, on geology.

Here, Darwin pauses once again to talk of "how complex and unexpected are the checks and relations between organic beings which have to struggle together in the same country" (p. 71). He begins his case not with reference to the exotic worlds of South America or Africa, but to "an estate in Staffordshire," where, he says, he had "ample means of investigation." The subject is once again a question provoked by a comparison, for it is only by noticing relationships that Darwin's prose (and science) can proceed—this time between a barren, untouched heath and another area "of exactly the same nature," that had, twenty-five years before, been enclosed. One should not underestimate the value here of Darwin's taking his example from the kind of world his readers would likely have been familiar with. He notes that though "of the same nature," the differences between the two areas had become unusual (not, he says, the sorts of differences that would develop normally even between areas with very different soils): "not only the proportional numbers of the heath-plants were wholly changed, but twelve species of plants (not counting grasses and carices) flourished in the plantations, which could not be found on the heath." There is a persistent charm, throughout the *Origin*, of a Darwin doing domestic experiments, gathering weedy mud in a cup, floating seeds in a little tank, counting, timing, noting, and that charm is part of this rather important (and slightly domesticated sequence). Darwin certainly took advantage of his ample opportunity to walk around, take notes, count plants, consider their relation to plants in the other area, and note differences. But he does not stop with plants. "The effects on the insects must have been still greater." He knows this not from counting the insects, of course, but from noting that "six insectivorous birds were very common in the plantations," and not to be found on the heath. On the heath he noted two or three other insectivorous birds, not to be found in the plantations—clearly, different insects, then.

Darwin spins out the connections, plants attract insects, and insects attract birds. There follows something like an exclamation: "how potent has been the effect of the introduction of a single tree, nothing whatever else having been done, with the exception that the land had been enclosed, so that cattle could not enter" (p. 71). Here then is the characteristic Darwinian pattern that begins with a wonder-inducing phenomenon and then chases down the explanation. He comes to realize what the problem in Staffordshire is by looking at an area "near Farnham, in Surrey," for there he came to understand "how important an element enclosure is." Again, as in South America, he comes to understand one area by thinking about it in relation to another, but now he is comfortably close to home.

How does all this work? Well, he noted near Farnham, where many areas had been enclosed, that on "distant hill-tops" "with a few clumps of old Scotch firs," over the last ten years of enclosures, "self-sown firs are now springing up in multitudes, so close together that all cannot live." Once again, "so much surprised," Darwin goes off to examine "hundreds of acres of the unenclosed heath." And across that whole wide expanse, surprised again, he finds not a single Scotch fir. Once again he must explain away the surprise.

"On looking closely between the stems of the heath," he finds

a multitude of seedling and little trees, which had been perpetually browsed down by the cattle. In one square yard, at a point some hundred yards distant from one of the old clumps, I counted thirty-two little trees; and one of them, judging from the rings of growth, had during the twenty-six years tried to raise its head above the stems of the heath, and had failed (p. 72)

because of the grazing of the cattle. It is hard not to use our own imaginations as we read this, and to see Darwin down on his knees in the grass, counting one by one the "little trees," and then, possibly still on his knees, scrupulously counting the rings of one of those "little trees." With such charming science, Darwin thus explains the surprise, and yet the effect is greater wonder: at the way in which the simple act of enclosure accounts for the growth of a Scotch fir forest, at the way the presence of a single tree changes the entire ecology of an area—cow, tree, bush, insect, bird!

But as if to create the experience of that dizzying infinite that always escapes full explanation, he does not stop there. If it is true that "cattle absolutely determine the existence of the Scotch fir... in several parts of the world insects determine the existence of cattle." It is an extravagant and, again, radically counter-intuitive argument (which he extends also from cattle to horses and dogs, as well), taking the form of paradox, which seems the only way to encounter the infinite complexities of nature—"plants and animals, most remote in the scale of nature, are bound together in a web of complex relations" (p. 73). The flies in Paraguay, which lay their eggs in the navels of these animals when first born, are the culprits that keep the larger animals from successfully breeding. Here, to get some richer sense of the world Darwin is evoking, it is necessary to quote even more extensively:

The increase of these flies, numerous as they are, must be habitually checked by some means, probably by birds. Hence, if certain insectivorous birds (whose numbers are probably regulated by hawks or beasts of prey) were to increase in Paraguay, the flies would decrease—then cattle and horses would become feral, and this would certainly greatly alter (as indeed I have observed in parts of South America) the vegetation: this again would largely affect the insects; and this, as we just have seen in Staffordshire, the insectivorous birds, and so onwards in ever-increasing circles of complexity. We began this series by insectivorous birds, and we have ended with them. *Not that in nature the relations can ever be as simple as this.* (p. 72–73, my emphasis).

It is comic. It is breathtaking. And it is awe-inspiring, all at once. We begin with surprise, explain why it shouldn't be surprising, and are breathlessly led through a sequence of connections that may be explained, but ultimately unknowable and yet more surprising. Darwin's is a Rube Goldberg sort of world, beyond the imagination of even that wonderful cartoonist, who, in rigging up various fantastic sequences of cause and effect, comically defied any normal sense of efficiency, and yet somehow managed to show that the job got done after all. What better metaphor for natural selection?

Darwin has this kind of ecological imagination, the recognition that the head bone's connected to the neck bone, and thus can see and connect the very large consequences that any small change in environment, any particular behavior, might have triggered, and continues to trigger at this very moment. But since nature is never this simple, we are destined to end our relationship to his world in that "entangled bank" (which turns up well before the book's famous last paragraph, in the chapter on "The Struggle for Existence," 74).

Again in that chapter, he talks about "beautiful diversity," about the almost miraculous way in which, "when an American forest is cut down, a very different vegetation springs up," but that somehow, "the trees now growing on the ancient Indian mounds . . . display the same beautiful diversity and proportion of kinds as in the surrounding virgin forests:

Throw up a handful of feathers, and all must fall to the ground according to definite laws; but how simple is this problem compared to the action and reaction of innumerable plants and animals which have determined, in the course of centuries, the proportional numbers and kinds of trees now growing on the old Indian ruins. (p. 75)

Is it mere eccentricity to be moved and awed by this vision of ordered significance within almost infinite forms of life, traveling through time and complexities of relationship impossible to capture—one can only imply them—in language? Here is a point that looks like an almost ultimate extension of Darwin's powers of "seeing," as he developed them on the *Beagle* voyage. Is it mere eccentricity to feel the excitement and energy of the life that surges from the visual into imagination and language, or to find value in and care for this world of puzzling, lovely flying feathers? Is it disenchanting to detect not merely struggle but multiple dependencies in Darwin's nature, to be dazzled by its constant motion, and to see, deeply implied beneath its visible surfaces, the endless complexities and intricate entanglements of a world stochastically organized, mysterious, and explicable at the same time? Is it merely Polyannaish to sense profound value in that intricacy of connection and that irregular beauty, or to wonder at the frightening power of those lawfully wild processes?

To achieve that vision, Darwin needed more than his extraordinary skills as observer and scientific theorist. He was a writer, perhaps in spite of himself—his metaphors and similes do what we might call extra-scientific work, even as they serve his science. Is it a poet who thought of that "handful of feathers" or a scientist? It is certainly an artist. The felt reality and the wonder of the world Darwin imaginatively evokes in the *Origin* depends upon an extraordinary creative imagination and the skills of a writer who read Wordsworth and Milton as well as de Candolle and Lyell.

Notes

1. "The Rhetoric of the 'Origin of Species,'" in Michael Ruse and Robert J. Richards (eds.), *The Cambridge Companion to the* "Origin of Species," (Cambridge: Cambridge University Press, 2009), 244.

2. As a scientist, Darwin cannot be enrolled among the "Victorian Sages," but in this book I am after those aspects of his work that are consistent with the Sages' strategies and commitments. The fundamental qualities of the Sage,

as John Holloway classically formulated them, are remarkably close to those aspects of Darwin's work on which I am focusing here. First, for the Sages "acquiring wisdom is somehow an opening of the eyes, making us see in our experience what we failed to see before." This is the central project of the *Origin*, as well. Second, "what gave their views life and meaning lay in the actual words of the original, in the Sage's own use of language, not in what can survive summarizing of their 'content.'" Of course, Darwin's summarizable meaning is profound and memorable, but I want to emphasize that aspect of Darwin that is not summarizable into his "ideas," and this, as for the Sages, lies "in the actual words of the original," which give his work yet another "life and meaning." Third, the Sage works "by quickening the reader to a new capacity for experience." We have seen this already in Darwin's *Beagle* prose and Holloway aptly for my argument quotes Conrad's purpose, which was, "above all, to make you *see*." *The Victorian Sage: Studies in Argument* (New York: W. W. Norton, 1953), 9–10.

3.   Here are a couple of typical early instances: "I could give many facts, showing how anxious bees are to save time," he notes at one point; "I could show by a long catalogue of facts, that parts which must be called important.... sometimes vary in the individuals of the same species."

4.   T. H. Huxley *Life and Letters* (MacMillan and Co., 1900), I, 183.

5.   See Robert J. Richards, *The Romantic Conception of Life* (Chicago: University of Chicago Press, 2002), 534–7, and his essay in *The Cambridge Companion to Darwin*, 100.

6.   It has long been understood that Darwin was much influenced by the Natural Theology that much of the *Origin* is devoted to displacing. The idea of nature as intelligently created is part of Darwin's initial approach to it: he needs to explain what looks like "workmanship" by demonstrating that it is not designed by an Intelligent Designer. And yet his imagination of nature is of a designed place, and he adapts many natural theological terms, not least "adaptation" and "contrivance" in his description of the way nature works. For the key scholarship in regard to Darwin's relation to natural theology, see Dov Ospovat, *The Development of Darwin's Theory: Naatural History, Natural Theology and Natural Selection, 1838-1859* (Cambridge: Cambridge University Press, 1981).

7.   "Several writers have misapprehended or objected to the term 'Natural Selection.' Some have even imagined that natural selection induces variability, whereas it implies only the preservation of such variations as arise and are beneficial to the being under its conditions of life. No one objects to agriculturalists speaking of the potent effects of man's selection; and in this case the individual differences given by nature, which man for some object selects,

must of necessity first occur. Others have objected that the term selection implies conscious choice in the animals, which become modified; and it has even been urged that, as plants have no volition, natural selection is not applicable to them! In the literal sense of the word, no doubt, natural selection is a false term; but who ever objected to chemists speaking of the elective affinities of the various elements?—and yet an acid cannot strictly be said to elect the base with which it in preference combines. It has been said that I speak of natural selection as an active power or Deity; but who objects to any author speaking of the attraction of gravity as ruling the movements of the planets? Every one knows what is meant and implied by such metaphorical expressions; and they are almost necessary for brevity. So again it is difficult to avoid personifying the word Nature; but I mean by Nature, only the aggregate action and product of many natural laws, and by laws the sequence of events as ascertained by us. With a little familiarity such superficial objections will be forgotten." Darwin inserted this paragraph into the third edition of the *Origin*, and it remained there through the 6th edition. It rigorously backs away from the implications that I have been pointing out, but those implications were there at the outset of Darwin's writing on the subject (versions of a very active natural selection appear in the early drafts—1842, 1844—of his argument, and vestiges of that personified process remain throughout later editions of the *Origin*).

8. Daniel Dennett, *Darwin's Dangerous Idea: Evolution and the Meanings of Life* (New York: Simon and Schuster, 1995), see esp., 48–60.

9. In the 1842 draft, for example, Darwin talks of "a being infinitely more sagacious than man (not an omniscient creator)," who, "during thousands of years were to select all the variations which tended towards certain ends"; who, Darwin then asks, "seeing . . . what foolish man has done in a few years, will deny an all-seeing being in thousands of years could effect (if the Creator chose to do)" (p. 36). It appears in much less metaphorical form in the 1844 volume, where Darwin expresses the same ideas: "how incomparably truer then would a race produced by the above rigid, steady, natural means of selection, excellently trained and perfectly adapted to its conditions, free from strains of blood and crosses, and continued during thousands of years, be compared with one produced by the feeble, capricious, misdirected and ill-adapted selection of man" (p. 94). The 1844 version is consistently less metaphorical than either the very short 1842 version or the *Origin* itself, and it seems to me that had it been published, its effects would have been less powerful than those of the *Origin*, or, most likely, the contemporary *Vestiges of the Natural History of Creation*.

10.   Paul H. Barrett, Donald Weinshank, and Timothy Gottleber, *A Concordance to Darwin's "Origin of Species," First Edition* (Ithaca: Cornell University Press, 1981).

11.   In *Angels and Ages: A Short Book About Darwin, Lincoln, and Modern Life* (New York: Knopf, 2009), Adam Gopnik, whose analysis of Darwin's writing is otherwise so insightful, argues that Darwin is not a metaphorical writer: "the remarkable thing about Darwin as a writer is not how skillfully he uses metaphor but how artfully he avoids it" (1494–500, kindle). It is true that Darwin's modest prose is as plain as he can make it, but every important concept in the book is metaphorical, not least "natural selection," and "sexual selection," not to speak of the metaphor of the "wedge."

12.   *Charles Darwin's Notebooks*, eds. Paul. H. Barrett, Peter J. Gautrey, Sandra Herbert, David Kohn, and Sydney Smith (Ithaca: Cornell University Press, 1987), "Notebook D", 135, pp. 375–6.

13.   In the sketchy 1842 version it was one sentence: "a thousand wedges are being forced into the oeconomy of Nature" (p. 37), and in 1844: "Nature may be compared to a surface, on which rest ten thousand sharp wedges touching each other and driven inwards by incessant blows" (p. 90).

14.   "In the universal struggle for existence, the right of the strongest eventually prevails; and the strength and durability of the race depends mainly on its prolificness, in which hybrids are acknowledged to be deficient." Charles Lyell, *The Principles of Geology* (Chicago: University of Chicago Press) II, 56. This is a facsimile of the 1832 first edition. It is notable that Lyell uses this phrase, which must have had some effect on Darwin, in order to prove that evolution is an impossible theory. He assumes that the struggle for existence will disallow hybrids from "prolificness" in breeding and will assist only the strongest members of each species to reproduce themselves. The assumption is that only through "hybrids" can evolution of one species into another be accomplished.

15.   "These different kinds of struggle, in Darwin's estimation, can be aligned according to a sliding scale of severity. Accordingly, the struggle will move from most to least intense: between individuals of the same variety of a species; between individuals of different varieties of the same species; between individuals of different species of the same genus; between species members of quite different types; and finally, between individuals and climate. These various and divergent meanings of struggle seem to have come from the two different sources for Darwin's concept: de Candolle, who proclaimed that all of nature was at war, and Malthus, who emphasized the consequences of dearth to whole populations. Today, we would say that struggle—granted

its metaphorical sense—properly occurs only between members of the same species to leave progeny. Adopting de Candolle's emphasis on the warlike aspects of struggle may have led Darwin to distinguish natural selection from sexual selection, which latter concerns not a death struggle for existence but males' struggling for mating opportunities." Robert J. Richards, "Darwin's Theory of Natural Selection and its Moral Purpose," *in Oxford Companion to the* "Origin of Species" (p. 63).

16.   Kropotkin cites this passage in *The Descent of Man*: "those communities, which included the greatest number of the most sympathetic members, would flourish best, and rear the greatest number of offspring." Charles Darwin, *The Descent of Man, and Selection in Relation to Sex*, 2nd. edn, 1879 (London: Penguin Books, 2004), 130 (Part 1, ch. 4).

17.   Jonathan Smith, *Charles Darwin and the Victorian Visual Imagination* (Cambridge: Cambridge University Press, 2009).

4

Surprise and Paradox: Darwin's Artful Legacy

In earlier chapters, I have tried to describe the development of Darwin's way of seeing and writing, and to emphasize elements quite different from the ones usually emphasized, the Darwin of "nature red in tooth and claw," the Darwin of Hardy's bleak world of mindless disasters, the Darwin whom Richard Dawkins so often wants to emphasize, in his bravura resistance to sentimentalizing nature.[1] For the most part, although I sympathize greatly with those who find such rhetoric excessive and in a way destructive to the cause of prosyletizing for the acceptance of Darwinian evolution by the culture at large, I find Dawkins' arguments—insofar as we are talking about Darwin's ideas—very reasonable. Nature *is* entirely indifferent to the things that humans value, it works blindly and by chance—a perfectly sensible inference from Darwin's theorizing, despite Darwin's repeated insistence that he does not believe in chance. "We cannot admit," complains Dawkins, "that things might be neither good nor evil, neither cruel nor kind, but simply callous—indifferent to all suffering, lacking all purpose."[2] In this respect, turning natural selection into an "algorithm" works.

But Darwin, the writer, also takes us somewhere else. The algorithmic Darwin is a product of modern interpretations—good ones, to be sure, and more than useful for scientific work, but not quite what Darwin wrote. For the rest of this book, I want to attend to one major element of what he did write—the element of surprise that issues in paradox. We have already seen that much of what he wrote was immediately startling, even when laid out in a quiet and businesslike language. The prose pattern foregrounds the startling nature of his subject and then, quietly, empirically, explains why it shouldn't be that startling once we think about it. Yet, we have seen, the prose manages, even in understatement, to make the world seem even more startling after explanation than before. Here I will be investigating this strategy and its effects as they figured importantly in Darwin's literary legacy. How did they work? What important Darwinian ideas depended upon them? Where do we find their traces among writers who followed him?

In the first chapter, I invoked an essay by Dwight Culler, in which he claimed that tracing the important influence of Darwin's ideas on literature doesn't "quite get at the heart of the problem" (p. 225). Those ideas, and their potentially dark implications, have been quite obvious to students of Thomas Hardy. (But in a later chapter I want to look at an important and different aspect of Hardy's work, which reflects not simply those ideas about a grimly indifferent universe, but a profound sense of life's abundance and resistance to formulas.) We can hear strange variations on the grimmer themes throughout Conrad's work, in the darkness of Central Africa, when Marlow, horrified, recognizes in the beating of distant drums the beating of his own heart, just as a horrified Darwin, encountering the Fuegians on their home grounds for the first time, could "hardly make himself believe that they are fellow creatures" (p. 213); or when, in Conrad's *Under Western Eyes*, the protagonist Razumov betrays the revolutionary Haldin, scrawling "evolution, not revolution." Darwin is obviously there, less malignantly, in Kipling's *Jungle Book* and *Kim*, but ferociously in American literature, in *The Call of the Wild*, in Dreiser's novels, and in the work of Frank Norris.

Let's say that all the news is not bleak, all the world not mindlessly and heartlessly cut-throat. Culler, as we have seen, points out surprisingly the comic form of Darwin's argument, the counter-intuitive way he sees the world. And partly because reading Darwin has always made me feel more, not less alive, always made me value life more, not dwell on its emptiness and meaninglessness, I want to think about the counter-intuitive comedy that is inherent in the very strategies of his prose. We have already seen how, building on a well-established Romantic tradition, his writing exhibits the most intense and focused attention to the particularities, richness, complexity, and almost universal unexpectedness of nature. Ironically, even the darkest echoes of Darwin's ideas—as I hope I have by now shown—are likely to fill the world with life, excitement, and strangeness, and fill art with new ways of seeing and shaping.

So, for example, when one hears the grinding of trees competing grimly against each other for space in the thick woods of Hardy's *Woodlanders*, it is not the idea of the struggle but the startling implication that the trees have sensibilities and voices that is most surprising and most moving. Yet more startling, Darwinian influences in Walter Pater, and in many others, produce a rarefied intensification of subjectivity, a questioning of the nature of self-hood, and a new exploration of value. Although I believe one can find these same qualities in Hardy and Conrad as well, they are most obvious in Pater's famous "Conclusion" to the *Renaissance*, which notoriously seeks its intensity in the texture of experience rather than in the hard reality beyond consciousness.

Despite Darwin's absolute commitment to "sharp and eager observation," to the objectivity of scientific work and the demands of material evidence, his theory transforms much of what was thought of as objectively real into the subjective. Darwin's deep materialism had the paradoxical effect of intensifying, for many, the focus on mind. That is, the solidity and impermeability of the material world left the only space for creativity, design, and meaning in the self, and that too, in Darwinian terms, is the product of

material processes. (I will be discussing some of Darwin's treatment of this question later on.) It's all a great paradox, but one can
see, reading Darwin carefully, that as he figures out how nature
and natural selection work, the virtuoso shaping of what seems like
the infinite, or the unknown, or the imperfect messiness of things
is Darwin's own brilliant theorizing.

The greatest paradox of Darwin's work, as Culler points out and
Daniel Dennett is always busy reminding us, is that the world's
design is not designed—it is the product of the absence of intelligence. It emerges (in an entangled way, of course) from the mindless
movements of nature. The implication is—Darwin doesn't have to
spell it out—that to account for the fact that all living things are
adapted to their niches, we have invented the idea of a designer.
That the world begins not with the fiat of an intelligent being but
through the slow hit and miss movements of nature is a reversal of
common sense; think about it for a moment and one realizes that
such a reversal entails the view that it is the mind itself, not nature,
that creates order. Darwin wasn't a philosopher and certainly not an
idealist: but his prose regularly finds ways to dramatize the mind's
power to make up order, all of which is almost burlesqued in the
second chapter of the *Origin*, in which he discusses the reality of
species. He talks of varieties that have been called species, species
that have been seen as varieties, and of a Mr Babbington, who "gives
251 species, whereas Mr. Bentham gives only 112,—a difference of
139 doubtful forms!" (p. 48), and the exclamation point is, of course,
Darwin's own. (He loves exclamation points!) And finally he seems
to sigh, "I was much struck how entirely vague and arbitrary is the
distinction between species and varieties" (p. 48). His point here is
not to provide a surer count of species and varieties but to lead us
out of the idea that species is even a real category (it is merely a
useful one, which allows us to impose an order on the world that
does not correspond to its reality).

Certainly no clear line of demarcation has as yet been drawn between species
and sub-species ... differences blend into each other in an insensible series;

and a series impresses the mind with the idea of an actual passage.... I look at the terms species, as one arbitrarily given for the sake of convenience to a set of individuals closely resembling each other, and that it does not essentially differ from the term variety, which is given to less distinct and more fluctuating forms. (pp. 51, 52)

This sharply defined and perceived reality turns out to be indistinct and fluctuating, like the earth Darwin describes in the *Journal of Researches*. The solidity and permanence of species, like that of the rocks, is an attribute of our perception rather than of what is perceived.[3]

What could be stranger than a book devoted to the "origin" of species that denies the reality of species (no less refuses to discuss true origins)? While by now, for students of Darwin and biologists, this is old news, it certainly was not old news in the last half of the nineteenth century: the idea was paradoxical and startling. Ironically, perhaps, Darwin's scientific materialism helped contribute to an intellectual climate in which idealism had become a powerful force, and in which study of the workings of the mind as a function of physiology was developing along many lines.[4] The mind, a product of physical nature, invests nature with a meaning and order that, intrinsically, it does not have.

It makes sense then that we can detect an elaboration of this kind of reversal in Pater, who claims that the aim of the critic is not, as for Matthew Arnold, to know the object as in itself it really is, but to know one's own impression as it really is. Kant had long ago taught everyone, except Arnold it seems, that the mind had no access to the thing in itself. So, on Darwin's back, Pater glides perilously close to solipsism and leaves us with a vision of the world in constant flux (recognizable in Darwin's *Journal of Researches*), from which value is extracted just by intensity of observation, by consciousness roused and startled. Experience, that key to empirical philosophy and to the development of early modern science, weaves and unweaves the self as the fluid world continually vanishes away.

Culler certainly didn't mean, nor do I, that Darwin's view of the world is "ha-ha" funny, or that it is always optimistic. Conrad's utilization of Darwinian surprise and paradox is hardly a big laugh, though it often manifests itself in a sort of dark humor as we consider the grotesques and the nightmares that inhabit so many of his stories. But funny or not, Darwin's rendering of the counter-intuitive world his investigations had uncovered is dazzlingly alive. The aesthetes and late-century "decadents" may seem a long way from the entirely domestic Darwin sitting in his famous study, staring at flowers, puttering in his garden, counting seeds, being as respectable as one man might reasonably be expected to be; but Gowan Dawson has recently shown how frequently Darwinian ideas and aesthetic writing, Swinburne as well as Oscar Wilde and Pater, appeared in the same journals, and how easily the aesthetes assimilated Darwin.[5] Of course, they took him where Darwin had no interest in going: into Wilde's upside down world, into his ironies about the decay of lying and criticism as art, and into the inward turn of so much of late nineteenth-century litera-ture. Consciousness does not inhere in the world, but in humans (and, to a lesser but important degree, in other organisms); matter is less trustworthy, or at least less apprehensible, than the play of consciousness; free indirect discourse or, then, stream of con-sciousness displaces the dominant realist mode of third-person omniscient narrative. "Mr Babbington" and "Mr Bentham," in their scientific taxonomy, might be seen as, after all, not really talking about nature but about their impressions of it, anticipating in that way Pater's impressionistic sense of the work of the critic.

I am not suggesting that Darwin was on the verge of solipsism. But his uncompromising naturalism, which prohibited him from finding explanations outside the workings of a recognizable nature, entailed, as we shall see, an intensified consciousness, a fine sensi-tivity to the way the mind worked and interacted with the visible (and sensible) world. As we have seen both in the *Journal of Researches* and in the *Origin*, he often, facing experiences of great and surprising intensity, tried to come to terms with matters that simple rationality would hitherto have left to the supernatural. In his rejection of natural theology, attempting to explain the "order" of the world through an intelligent designer, and in his implicit critique of the tradition of identifying species in which he had begun his scientific life, Darwin exposes the way the mind imposes itself on nature. Seeking to get beyond mind to the reality itself, what Lamarck had called the "natural order" rather than the scientific one,[6] Darwin discovered a world in which the mind was the producer of meaning after all; it did not inhere in the nature of things. Such a vision, with its implicit emphasis on the creative activity of mind, becomes part of the worldview of much modernist literature.[7]

The form of Darwin's thought and language can be detected in some unlikely and ostensibly unscientific places, where subjectivity and aesthetic value displace the quest for "truth" and the full look at the worst. But perhaps "displace" is the wrong word; I would prefer "supplement." Darwin's quest for truth was built on qualities that are essential to both scientist and writer. So the Darwin that I am emphasizing here does not so much contradict the Dawkinsian scientist as bring, in his work of registering the way the world is and reasoning through its strangeness, a sense of wonder and a sense of value that might be shared by everyone.

In a recently published essay, Robert J. Richards has argued for the importance in Darwin's work of consciousness itself, and the compatibility of his theory with questions of purposiveness and value of the sort that Dawkins insists natural selection banishes from the world. "Little wonder," Richards says,

that we still think of organisms and their parts in terms of purposes, of ends. Only mind can hold the past and present together and thus come to understand the activities of organisms in terms of the aims that they display. A flower may achieve its goal in producing the nectar that attracts insects in order that its pollen may be spread to other flowers, but only a mind can come to understand the presence of nectar in the flower in relation to its typical end. What modern biologist would deny that the lens of the eye is designed for the purpose of casting a coherent image on the retina? We now understand such designs as the result ultimately of natural selection...we now conceive mind as if it were a natural selection device. That is, in the effort to solve any problem, we imaginatively fling out variations, possible solutions, until one seems to work—or work sufficiently well for the moment. This kind of blind variation and selective retention grounds most of our inventive efforts at thinking through problems. Darwin may have been moved to construe natural selection in terms of mind, because mind, in fact, works like natural selection. For anyone who investigates the process by which Darwin came to construct his theories is well aware that he proceeded by imaginatively trying out this variation and that variation, running down blind allies until some one of those avenues advanced the solutions he sought. He may well have been dimly aware of his own typical activity in this respect. Thus at a very deep level his own mind may have served as the template for understanding nature. The world remains an enchanted place, though the source of that enchantment may be other than often supposed.[8]

Thus, if the turn toward tragic endings in late-century novels was connected with Darwin's mindless, chance-driven world, so too was the "inward turn." And the possibility of finding in this Darwinian world the joy and wonder of the mind at work as it discovers (and creates) startling connections in nature gives considerable support to Culler's contention. The metaphorical natural selection making choices and "tending to" each organism, was essential to Darwin's initial conception of it because, as Richards has consistently argued, his original model assumed "a mind in nature."

Darwin, of course, understood how difficult it would be for contemporaries to absorb his way of thinking and seeing, and the upside down and fluctuating world he described. So near the end of the *Origin*, anxious about the reception he might receive, Darwin confesses that he by no means expects "to convince experienced naturalists whose minds are stocked with a multitude of facts all viewed, during a long course of years, from a point of view directly opposite to mine." But, he says, "I look with confidence to the future, to young and rising naturalists, who will be able to view both sides of the question with impartiality" (p. 481). It was a modernist leap that Darwin anticipated, and to grasp fully the art of Darwin's prose requires a modernist shift in point of view: Darwinian things as they are can only be perceived (or "created") by way of a change of perspective that challenges what seemed like common sense. Richards is right about the importance of mind, the Darwinian mind, in our understanding of the world governed by natural selection. Darwin's new sublime is not so much outside in the wonders of nature he so much admired and felt, as inside, in the power of mind to imagine beyond what it sees. The shift partly cuts the essence out of a language that had attempted to describe a "natural system" and had been thought to reflect Nature as it is.

As Gillian Beer has classically shown, the language Darwin used and resisted, which is also our language, always implies agency, and is in addition intrinsically anthropocentric—two qualities that Darwin labored hard to reject. Nouns imply firm boundaries and absolutely distinguishable entities. It takes a lot of language to overcome the implications of language, and this is one of the reasons that Darwin's writing so often took the form of paradox, the form that is at the heart of the aesthetic turn at the end of the nineteenth century and in modernist literature. "Natural history," said G. H. Lewes in 1860, "is full of paradoxes."[9]

Darwin had not only to make the rock-solid scientific case but also (against his awareness of how hard this would be), to change our sense of probability, and ultimately to change sensibilities. "I remember too well," he wrote to a skeptical reader, "how many long years my conversion took," and that word "conversion" cannot be a casual one. Darwin's work revises the world by revising our language, by investing it with mind, through metaphor and personal intrusions, and making us feel the inadequacy of what we had thought were our common-sense intuitions of how things are.

To achieve that, of course, he depended heavily on the quality we have investigated most fully in the discussion of the *Journal of*

Researches, that is, the only exceptional talent he unequivocally attributed to himself in his *Autobiography*—the power of observation, which entailed questioning and thinking as much as physical perception. As he wrote to Wallace in 1857, "I am a firm believer, that without speculation there is no good & original observation." (*CCD*, vol. 6, p. 514). Darwin turns our world upside down as his mind plays about his observations, and at the same time, constructing a counter-intuitive argument, tries to make common sense of the new vision.

Seeing, as one member of the future generation, W. K. Clifford, put it, leads us into paradox. In a famous essay on "The Philosophy of the Pure Sciences," Clifford begins by describing what he and everyone else seems to see upon entering the auditorium, and then dramatically reverses himself—"And yet," he says, "I think we shall find on a little reflection that none of these statements can by any possibility have been strictly true."[10] Clifford reminds us that we create what we see beyond the limits that the structure of our own eyes shape and determine. "In every sensation there is, besides the actual message, something that we imagine and add to the message...not the whole of sensation is immediate experience" (p. 309). The act of seeing is itself an act of imagination, although we may be deceived into thinking that all we "see" is really there. The auditorium that we are confidently looking at is a construction of our mind inferred from the limits of our visual powers. Unlike the rhetorically more tempered Darwin, Clifford showily performed paradox in his wonderful, dramatic, and, as William James called them, too "robustious" essays and lectures. But those essays look back toward Darwin's counter-intuitive vision, while they point forward to the more demonstratively paradoxical modes of the *fin de siècle* and Oscar Wilde.

Clifford's theory of mind, though he died far too young to develop the theory completely, corresponds valuably to the idea that a Darwinian view of the world helped inspire the "inward turn." He faces directly the question that his point, above, seemed to imply, that is, "if we have to admit that there is no real message from without, *all* the sciences will become pure sciences, all knowledge will be *a priori* knowledge; and we may construct the universe by sitting down and thinking about it" (p. 338). In Clifford's way of talking about it, empiricism, ironically, leads to something like Pater's near solipsism, which in its turn denies the necessity of experience after all. This is no place to wander into the fogs of this kind of philosophy, and certainly Darwin never allowed himself to be sufficiently entangled to challenge the power and the necessity of objective observation and experience. But the borderline here, once one recognizes that consciousness, because it is the product of limited material conditions, does work that can never adequately affirm the real existence of matter, is an area in which much Victorian thinking, writing, and art are conducted. For an artist, there is no escaping, as a consequence of this philosophical muddle, the crucial importance of consciousness itself, and the recognition that whatever the outside world is *really* like, that hard material "unknown" of Herbert Spencer's theorizing, we respond to the activities of consciousness. The action is *inside*.

For Clifford, the most important condition of mental development was resistance to conventionalities; what was required was "plasticity" of mind and he concludes his essay with a determined paradox in the mode of late-century iconoclasm: "It is not right to be proper." Beginning an essay "On Some of the Conditions of Mental Development," with a startling assertion, Clifford writes: "If you will carefully consider what it is that you have done most often during this day, I think you can hardly avoid being drawn to this conclusion: that you have really done nothing else from morning to might but *change your mind*" (p. 80). Such attitudes, paradoxical, anti-conventional, counter-intuitive, are marks as well of Darwin's thought and of his prose; they might also be understood as a literary transition from Victorianism to modernism, borne on the back of science itself. If Darwin does not explore philosophically the complexities of empiricist theory, he nevertheless produces a world that forces later writers to think about them.

We have seen that Darwin had already broken through conventions of perception, to observe connections not literally visible to the eye, to infer history from static fact and movement from apparent stability. While taxonomy seemed to depend upon the way organisms shared characteristics, nothing was more valuable to Darwin's general theory than the singular, the aberrant, the anomalous, the exception, and the vestige. "Individual difference, though of small interest to the systematist," says Darwin, is "of high importance for us, as being the first step towards such slight varieties as are barely thought worth recording in works on natural history" (p. 51). *His* science entailed a reversal, finding that it is *not* what is useful to organisms that allows us to understand their taxonomic place, for the useful will have been shaped by natural selection; rather, our best indication of genealogy is what is *not* useful and thus untouched by natural selection and need. One might notice here that just as the art of nineteenth-century decadence aspired to reject the useful for the beautiful, so the epistemology of Darwinian science sought the useless as the most important signs of our history. "Art for Art's Sake" could be recognized as a self-conscious rebellion against utilitarian readings of life and a rebellion against the new bourgeois/industrial culture. It did considerable social work in it's denial of the relevance of society to art. This pattern of celebrating "Art" makes an ironic counterpart to the movement of Darwinian thought, for Darwin's science requires recognition that it is precisely *not* common sense (Huxley had called science "organized common sense"), not habit, but a trip through the looking glass. One gets to the truth by looking at what's not useful.

From this perspective, we can see even more clearly why Darwin's entire theory and all of its details, however soberly registered, amount to a giant paradox. Consider the conditions of Darwin's world. What is stable is in motion; what is enormous depends upon minutiae; what seems peaceful is at war; struggle is often mutual dependency; lowly worms create the large green expanses of England; six thousand years is no time at all; "species" is an arbitrary term, not really much different from "variety"; if unchecked by natural selection, even slow-breeding elephants would entirely cover the earth within five centuries; there are woodpeckers living where not a tree grows; there are web-footed birds that never go near the water; we are related physiologically to all living things, not only apes but barnacles and spiders. "We behold the face of nature bright with gladness... we forget that the birds which are idly singing round us mostly live on insects or seeds, and are thus constantly destroying life" (p. 61). The world of brilliant adaptations is moved not by a creative intelligence but by "unknown laws of nature," and in nature itself the only intelligence is that of organisms, and most obviously and particularly humans. "There seems to be no more design in the variability of organic beings and in the action of natural selection, than in the course which the wind blows"(*Autobiography*, 87). In earlier chapters I have offered yet more examples to show that this breathtaking litany could go on. Darwin's world emerges strange, unpredictable, sometimes comically perverse, sometimes awesome and scary. Alice, once through the looking glass, notes that "what could be seen from the old room was quite common and uninteresting, but that all the rest was as different as possible."[11]

In many of its aspects, this Darwinian world, affirming the importance of what had seemed unimportant, moving gradually through history to build to important changes, is consonant with the work of Victorian realist fiction. But particularly, as Jonathan Smith has pointed out, in Darwin's books about botany, there is something about Darwin's preoccupation with the ordinary that extends beyond realism. Darwin's preoccupation with the aberrant and the unique and the grotesque (as traces or indicators of larger movements and significances) extends in the botany books to the study of plants that are truly, as Smith puts it, "macabre,"

plants that trapped, killed and ate insects;... bizarrely ornamented, gaudily colored, fantastically-structured flowers that lured insects into unwittingly effecting cross-fertilization; and... species with various sexual forms and multiple reproductive possibilities. While these works also embodied a realistic spirit, the literature of the day that most looked like Darwin's botany was not the realistic novel of Trollope and Eliot, but the "sensation fiction" that flourished in those same decades, the thrilling novels full of shocking crimes and illicit sexuality by Wilkie Collins, Mary Braddon, and others.[12]

This grotesque and apparently aberrant aspect of Darwin's work was well recognized, as Smith shows, by his contemporaries. Hard as Darwin and others worked to keep Darwin within the frame of Victorian normality and respectability, his work constantly broke down those defenses.

Poets have always, in their metaphorical visions, had something of this perverse capacity to see things from a different point of view—"negative capability" is the label of choice. Let me invoke one of Tennyson's short, famous lines to give a sense of where I want to go with this idea—not as though it were directly influenced by Darwin, but because it exemplifies an aesthetic power reasonably associated with Darwin's writing. As Tennyson's famous "eagle" flies, "The wrinkled sea beneath him crawls." Because the poem is so conveniently short, it is ubiquitously anthologized, but for the same reason it is easy to lose sight of what a stunning vision—before the age of airplanes and moon walks—this is. As Darwin was to place us inside the consciousness of the female Argus pheasant, Tennyson gives us the eyes of an eagle observing, from the enormous heights at which it soars, what Matthew Arnold from much another more human perspective called the "unplumb'd salt, estranging sea." And where does the sublimity of this image lie?—not in the sea but in the eagle's perception of it. It is that perception, imagined and interpreted by the poet, that turns the sea, that traditional mythical site of birth and death, into something like a bedsheet or an old shirt.

Although at the moment of Tennyson's writing of the poem, Darwin was probably being made seasick by those wrinkles, the image serves as the kind of shift of perspective towards which Darwin was to labor. Emerging from the literary and scientific culture that precipitated Darwin's work, it implies one of Darwin's great achievements—the imaginative power to think beyond the human; moreover, it does so in the double movement characteristic of Darwin's writing, the radical juxtaposition of two incompatible conditions—the vast and the domestic. The sublime re-emerges from this juxtaposition by way of the extraordinary consciousness that is capable of holding them together.

The famous rejection of William Paley's natural theology establishes the form: Darwin shifts from an anthropocentric perspective to see the world from the eagle's point of view and understand that the eagle's eye view is as valid and valuable as the human's. He must have learned something of this strategy from Charles Lyell, who, in a wonderful passage to which I referred in the first chapter, from the first volume of *Principles of Geology*, reminds readers of how constrained their understanding is by the limits of their merely human perspective.

If we may be allowed so far to indulge the imagination, as to suppose a being, entirely confined to the nether world—'some dusky, melancholy sprite'—like Umbriel, who could 'flit on sooty pinions to the central earth,' but who was never permitted to 'sully the fair face of light,' and emerge into the regions of water and of air; and if this being should busy himself in investigating the structure of the globe, he might frame theories the exact converse of those usually adopted by human philosophers."[13]

I have to be careful here not to allow Lyell's wonderfully cultivated prose to upstage Darwin's, yet Lyell's alertness to the constraints of perspective is central to the scientific tradition that Darwin entered, and it is worth noting that Tennyson knew and was much influenced by Lyell's reading of the earth (it is, of course, Lyell, rather than Darwin, who lies behind "nature red in tooth and claw"), and for all three writers, we can recognize, as Gillian Beer has taught us, that Milton figured importantly.[14] George Eliot evokes Uriel in attempting to find a perspective equal to the complex interdependencies of relationships in *Middlemarch*. Lyell evokes Umbriel to correct our perspectives. Tennyson sees with eagle eyes. And Darwin, in one of the brilliant sequences of the *Origin*, invokes those eyes to "stagger" us (his word) (p. 204) into a recognition that sublime vision can grow from Lyellian gradualist causes. Think about how the world changes as our point of vantage changes. Think about the astonishing possibility that aquiline vision is simply a natural extension of the first light-sensitive tissue in some lower organism. Darwin, like Tennyson, thrusts us into the sublime along pathways of domesticity.

This Darwinian conjoining of the domestic with awesome vastnesses is linked to the shift of attention from the reality of nature, no longer laden with meaning and design, to the consciousness of humans. Although it is true that much of Darwin's argument in *The Descent of Man* depends on his demonstration that the mind of man is continuous with the consciousness of lower organisms, it is for him, as for us, the human consciousness that shapes the world. It takes human consciousness to recognize and value consciousness in other species[15]—although it is taking us far too long. The eagle, after all, has intelligence; and animal intelligence, in Darwin's reading, is at the basis of the development of both aesthetic and moral sensibilities in humans. But self-consciousness seems still to be exclusively human; the human capacity to "see" in the full sense I've been exploring, in effect half creates that world. The true sublime is not the sea, or the mountains whose histories Darwin wonderfully explains and which leave him awestruck, but the narrative imagination that sees the world in a grain of sand and that fills it with a meaning that, in its raw material thinginess, it doesn't have.

We have seen that in Darwin's world, everything "means"; everything evokes a history and entanglements of relationship. Darwinian explanation is necessarily narrative, producing hypotheses about every fact; and one might say that he turns the *Origin* into a series of convincing *Just-so* stories, some of which we have already considered, stories like those embedded in the famous tree-like diagram in the chapter on natural selection (see Figure 3.2 on page 105). They are thought experiments: this is what *might* have happened. This is the most probable of possible stories.

I want to consider here one of the most amazing of these *Just-so* stories in Darwin's work, not from the *Origin*, but from the *Descent of Man*. The pattern is the same; here wonder at the astonishing beauty and form of the tail of the Argus pheasant, and then an extended and arduous explanation of how the form developed—always actualist and probabilist—and then a conclusion as stunning and wonderful as the tail itself (see figure 4.1). The effect is—or might well be—not only a sense that the naturalistic process is as amazing and wonderful as the tail feathers themselves, but an almost awed recognition of the brilliance of the storytelling, or the mind that conceived such an explanatory narrative.

Here too Darwin writes as though he recognizes that he can never produce the smoking gun; there are far too many unknowns and multiplicities. He turns once again to probable narrative, which, as ever, privileges the empirical and the rational over the imagination, but surely it is imagination that allows Darwin to puzzle out how it would be possible for the Argus pheasant to

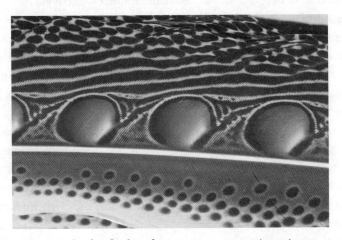

Figure 3. Detail of a feather from a great argus Argusianus argus. ©The Natural History Museum/Alamy.

develop so striking a tail without supernatural aid, that is, by the gradual incremental working of natural selection. The theory with which Darwin works to do this job, however, is sexual selection, not natural selection; nevertheless, the process remains entirely natural, unaided by any consciousness capable of designing the markings on the tail even though the tail certainly looks designed. Part of the startling nature of the extraordinary Argus feathering is that it appeals immediately to the human aesthetic sense, and the common-sense reaction to such a beautiful object is to credit it as designed. It would seem, then, that either the Argus pheasants were somehow capable of creating, or at least planning, human art, or the feathers were designed to satisfy humans rather than pheasants. We know that both of these possibilities are excluded from Darwin's theory, so that he had to try to figure out how to explain this phenomenon without invoking an intelligent designer, and without allowing the possibility that natural selection was producing a phenomenon meant to benefit not its producers but others. To accomplish this very difficult end, he needed to find a language that could displace the assumptions that design implies an intelligent designer, with a counter-intuitive explanation that would have to reveal an unintelligent designer—and a very intelligent explainer.

Gillian Beer has already shown that through the theory of sexual selection, Darwin reintroduces into the processes of nature an intelligence—expelled by the theory of "natural selection"—but not the intelligence of a supreme designer deliberately aiming at particular ends, only the more limited intelligence of the organisms who end up choosing but not creating the designs they like best. And these choosers, among the birds at least, turn out to be almost exclusively female. Female choice forms the foundation of birds' dimorphism, as Darwin discusses the question, and there have been strong and convincing arguments that Darwin does not transcend the limits of his culture's sexism with his emphasis on female choice.[16] Whether or not he did, in his consideration of such things as the feathering of the Argus pheasant, he was still stuck with the problem of those wonder-inducing ball-and-socket ocelli of the Argus pheasant, which have, as he says, "excited the admiration of many experienced artists."

Here again the basic Darwinian strategy:

That these ornaments should have been formed through the selection of many successive variations, not one of which was originally intended to produce the ball-and-socket effect, seems as incredible, as that one of Raphael's Madonnas should have been formed by the selection of chance daubs of paint made by a long succession of young artists, not one of whom intended at first to draw the human figure.[17]

The problem is how to account naturalistically for unintentional traits that seem intelligently designed, and we have a model for that in Darwin's brilliant handling of degrees of sightedness in his discussion of the eye in the *Origin*. There, echoing and overturning Paley's discussion of the eye in *Natural Theology*, he builds a story by showing that it is wrong to assume that keen sight like the eagle's could not have developed through intermediate stages of lesser and even very weak sight.

The trick then was to show that there are extant organisms that show every degree of gradation between the mole's sight and the eagle's. So, already convinced that natural selection works by way of "insensible gradations," he imagines, or finds, such gradations in the Argus pheasant's feathers (note how the world Darwin describes remains always just on the edge of being describable because of its complexity, infinitude, gradualness, or the unknown nature of historical causes. To borrow a word from Pater, there is something diaphanous in his prose, even in his most objective and logically directed language. He is always seeking the finest distinctions, the most imperceptible gradation.) Not having living examples of birds with intermediate grades of ball-and-socket ocelli, he turns existing feathers into a story, putting extraordinary realistic, almost three-dimensional images in significant relation to other, less obviously "intentional" markings on the feathers. Meticulously analyzing those relationships, Darwin shows "that a gradation is at least possible from a mere spot to a finished ball-and-socket ocellus" (II, ch. xiv, p. 142).

The subtlety of Darwin's argument here, built on minute attention to every nuance of shading in the feather, is difficult to reproduce without pages of quotation, and I confess that not all of the prose, so determinedly precise, is thrilling as prose, although its project and its effect are startling. There is evidence that the Argus pheasant wasn't always so showy, and that the current remarkable design in the feathers is a product of time. He then shows that what might seem mere random spots develop by way of female choice over many generations into the astonishingly beautiful design of the mature male Argus pheasant. The feather he sees at the moment reveals its history, and the transitions in space he sees becomes a story of transitions in time. Seeing spots on the Argus feather, Darwin sees movement as well. The feather is not locked into a particular moment, but is the image of transformation in motion, as Darwin traces the transition from spot to line to ocellus. Thus, the astonishing beauty that evokes the admiration of experienced artists is shown to have emerged from some simple spots. Yet ordinary spots, under the pressure of female desire, over long periods of time, transform into the spectacularly formed ocelli, and the effect of such process is not a diminishment of awe and admiration but a Darwinian intensification of it.

The Darwinian sublime is particularly counter-intuitive in that what is ultimately awesome and hard for the mind to comprehend is just the accumulation of materials not at all awesome and entirely easy to comprehend. It is that odd sublimity that Darwinian prose, with its repeated double movements, persistently creates. That the ordinary can achieve such extraordinary developments as the ball-and-socket occelli is the ultimate Darwinian wonder—and art. Although Darwin's detailed discussion is difficult to follow without an adequate image of the feather, it is worth looking at a piece of it to see how the prose works to transform the feather into a remarkable narrative, built on the most precise examination of it as a "rastro." I excerpt here a short, exemplary passage, and I suggest that you try to follow the argument by a close look at the image of the feather (see Figure 4.1). Trying that will suggest how much work beyond mere vision went into Darwin's seeing and into his prose:

Between one of the elliptic ornaments and perfect ball-and-socket ocellus, the gradation is so perfect that it is scarcely possible to decide when the latter term ought to be used. The passage from the one into the other is effected by the elongation and greater curvature in opposite directions of the lower black mark . . . , and more especially of the upper one, together with the contraction of the elongated sub-triangular or narrow mark, so that at last these three marks become confluent, forming an irregular elliptic ring. This ring is gradually rendered more and more circular and regular, increasing at the same time in diameter.

It takes Darwin eleven pages and four illustrations to produce a fully satisfying and convincing description that is also a narrative. It requires characteristic Darwinian attention to the subtlest of transitions (as there is no clear point at which a variety can be identified as a species, so here there is no clear point at which "the elliptic ornaments" become "ball-and-socket ocelli"). Darwin's world is dizzying in its refusal of essences and unequivocal identities; everything gets involved in the subtlest of movement in time and space—things pass into each other. Something of the infinitude and the unknown lingers around each of his stories, not because they fail to be pictorially precise, but because what they are after is ultimately "imperceptible." But the imperceptible finally does issue in something as clear and beautiful as the Argus pheasant ocelli: "Thus," concludes Darwin, "almost every minute detail in the shape and colouring of the ball-and-socket ocelli can be shewn to follow from gradual changes in the elliptic ornaments; and the development of the latter can be traced by equally small steps from the union of two almost simple spots" (II, ch. xiv 148).

Modestly and cautiously, Darwin concedes that he has only told a story: "the stages in the development exhibited by the feathers on the same bird, do not at all necessarily shew [sic] us the steps passed through by the extinct progenitors of the species; but they probably give us the clue to the actual steps, and they at least prove to demonstration that a gradation is possible" (II, ch. xiv, 150). A story is possible, even necessary, and it is a better story than anyone else has ever told about those feathers. Falling back on the implicit absolute reliance on the trustworthiness of causes now in operation, Darwin infers a story from the different textures and the graduated order of the marks: "Thus we can understand—and in no other way, as it seems to me—the present condition and origin of the ornaments on the wing-feathers of the Argus pheasant" (p. 151). The story shifts from the tentative ("we don't know, but we can see that this is possible") to the fully confident ("and in no other way").

Here are all the basic elements of Darwin's skills and vision: that brilliant power of observation that allows him to focus, nine pages' worth, on the small, sometimes random seeming, marks that accompany the ocelli; that capacity for inference and argument, along with his refusal of the idea that anything in nature is purposeless and the result of chance; that sense, then, that every object has a meaning; that power of reason, which is really a kind of imagination, that allows Darwin to detect history in stasis and to see transitions over long series of insensible gradations from the crude to the almost indescribably fine; and finally, that characteristic modesty that allows for the possibility that his own narrative is not completely correct. All of this is accompanied by a tenacious and even fierce adherence to his fundamental hypothesis.

It is an intellectual tour de force, the spelling out over pages of scientific tract, of an enormous paradox—art that is not artful, the beautiful without the artist. Or rather, the artist is concerned not with the art, as we see it in the feather, but with the possibilities of mating. The force of Darwin's story—as it was, for example, in his recognizing the quartz eroded from the top of huge peaks 45 miles distant from their mountain source—is once again not only in the natural object—here the ball and socket ocelli—but in the observer's powers to read its history, the imagination that connects large and small through space and time, that reads millennia into spots, stripes, and balls. Everything leaves its traces, has its history.

In the literary world blossoming around Darwin, there is one most obvious place to look for this peculiar kind of startled recognition issuing in fascinating storytelling and celebration of the inquiring mind: the very Victorian genre, the detective novel, echoed and technologized in our contemporary television police shows that spend half of camera time in labs where the slightest clue, a hair, a pin, a shred of cloth, mud caught in shoes, a daub of paint—all become means to telling the story of the crime and identifying the culprit. I keep asking myself as I watch those scenes in forensic science labs with test tubes and tweezers and high-tech machines, why I am watching this. What is it that virtually guarantees success to any narrative that can show a detecting mind winnowing meaning out of large collections of fact? Sergeant Bucket in *Bleak House* had already evoked Dickens' admiration. Wilkie Collins was already well at work on this sort of tracing the "rastro" in complex, famous sensation mysteries like *The Moonstone* or *The Woman in White*. But Sherlock Holmes was the figure who most captured the public imagination. I am not claiming, of course, that Darwin was the direct and major influence on the development of the detective novel; obviously, Dickens, Collins, and Edgar Allen Poe were inventing detectives before the *Origin* was published. Ronald Thomas, though also evoking Darwin, writes that "Sensation novels may be seen as domesticating and integrating all these historical investigations—of social class, of political legislation, of biological inheritance."[18] Obviously, a number of factors besides Darwinian biology went into the sudden emergence of this sub-genre, but there is no question that Darwin was enormously important in creating the intellectual and popular climate that could feel the pressure of history not only in objects but in their own bodies, and that could celebrate a new kind of intellectual power (Thomas associates this with the emergence of a professional class). Before Darwin, the development of comparative anatomy, especially with the spreading fame of Cuvier and Richard Owen, helped make the idea of being able to reconstruct the past a matter of wide popular knowledge, and scientists could become kinds of celebrity. But Darwin's *way* of looking at the world, which in fact produced a real exuberance at the possibility of finding things out that might previously have seemed unknowable, played a large role in the developing cultural enthusiasm for the genre. All detectives are on the trail of the rastro.

With Conan Doyle, one almost finds the smoking gun of connection to Darwin. In a 1908 book, *Through the Magic Door*, designed to induce young people into intellectual work, Holmes's creator takes Darwin as a model, noting how reading *The Voyage of the Beagle* revealed to him immediately the amazing quality of Darwin's mind:

Any discerning eye must have detected long before the "Origin of Species" appeared, simply on the strength of this book of travel, that a brain of the first order, united with many rare qualities of character, had arisen. Never was there a more comprehensive mind. Nothing was too small and nothing too great for its alert observation.... How a youth of Darwin's age...could have... acquired such a mass of information fills one with...wonder."[19]

The qualities Doyle, admires in Darwin are embodied in Holmes, who captured the public imagination because he was so Darwinian, so startling in his power to juxtapose and make a story of unlikely phenomena and the facts of ordinary life. We first meet Holmes in a story in which Watson discovers an essay that Holmes has written about "how much an observant man might learn by an accurate and systematic examination of all that came in his way":

From a drop of water...a logician could infer the possibility of an Atlantic or a Niagara without having seen or heard of one or the other. So all life is a great chain, the nature of which is known whenever we are shown a single link of it.

Like all other arts, the Science of Deduction and Analysis is one which can only be acquired by long and patient study, nor is life long enough to allow any mortal to attain the highest possible perfection in it. Before turning to those moral and mental aspects of the matter which present the greatest difficulty, let the inquirer begin by mastering more elementary problems. Let him, on meeting a fellow-mortal, learn at a glance to distinguish the history of the man, and the trade or profession to which he belongs. Puerile as such an exercise may seem, it sharpens the faculties of observation, and teaches one where to look and what to look for. By a man's finger-nails, by his coat-sleeve, by his boots, by his trouser-knees, by the callosities of his forefinger and thumb, by his expression, by his shirt-cuffs—by each of these things a man's calling is plainly revealed. That all united should fail to enlighten the competent inquirer in any case is almost inconceivable.[20]

"What ineffable twaddle" responds Dr Watson. Precisely the sort of response Darwin feared (and got from John Herschel's initial reported response to the idea of natural selection—"the law of higgledy piggledy"). But Holmes *does* demonstrate that everything is connected, and to the bewildered Watson's amazement reads Watson's life off a gold watch that Watson carried. One might regard Watson as the figure who plays the role that Darwin interpolated into his books: the wonderer and the doubter who has to be convinced by Holmesian or Darwinian demonstration, that the wonder is rationally explicable. Watson, like Darwin himself, plays the role of the audience itself, and the rhetorical work of such a character is done when he is finally convinced (as Darwin hoped the audience would be) by the scientist/detective's careful explanation. Watson charmingly accedes to Holmes's counter-intuitive brilliance—"Nothing was too small and nothing too great for [his] alert observation." Holmes can recognize that everything "means"; we recognize that he has the kinds of powers that Doyle had attributed to Darwin.

Playing with the counter-intuitive—formally a paradox—is, of course, playing with fire; the kind of ambiguous and startling fire that marks modernist literature's ironic reversal of Victorian values. While the directions *fin de siècle* writers took feel to be at odds with Darwin's sense of personal life and decorum, Clifford, clearly not a member of any aesthetes' circle, was already there. From a counter-intuitive science he attacks mere propriety and habit, as Pater was doing at the same time if in very different ways. Dawson has recently shown that Victorians themselves made the connection, and that Darwin—whose work makes genealogy and thus also sexuality the primary object of his attention—defended his respectability against the connection; but even the quickest look at Pater and Wilde makes clear that Darwin's way of seeing lay behind aestheticism as much as it did behind bleak naturalism and the detective novel.

Who more than Walter Pater, among the Victorians, emphasizes the ephemeral nature of things and shifts perspective to the inside, to sensibility? We know that Pater thought much about the development theory, although he saw Darwin's emphasis on change as only the most influential modern manifestation of a tradition going back to Heraclitus and developed in Hegel. He notes of Darwin in *Plato and Platonism*, that for him, "'type' itself properly *is* not but is only always *becoming*" (this, we have seen is a very accurate summation of Darwin's point about "species"), and that "the idea of development ... is at last invading one by one, as the secret of their explanation, all the products of mind, the very mind itself, the abstract reason; our certainty, for instance, that two and two makes four."[21] "Becoming" echoes an idealist and continental philosophical tradition, but Darwin, in particular with his work in *The Descent of Man*, and Spencer, were forcing the idea of development into the very activities of mind. And who if not Darwin lies behind the famous opening to Pater's essay on Coleridge: "Nature, which by one law of development evolves ideas, hypotheses, modes of inward life, and represses them in turn, has ... provided that the earlier growth should propel its fibres into the latter, and so transmit the whole of its forces in unbroken continuity."[22] In the essay, Pater invokes "those sciences" that "reveal types of life evanescing into each other by inexpressible refinements of change" (p. 66). For Darwin, "differences blend into each other in an insensible series; and a series impresses the mind with the idea of an actual passage" (p.51). Pater takes the next step: "Things pass into their opposite."

Pater's prose might be thought of as a set of experiments in developing the Darwinian challenge to devise a language that resists its own nature. His famous hostility to formulae and essences and insistence on the relativity of knowledge, and his recurrence to "inexpressible refinements of change," as what might be taken as a version of Darwin's "imperceptible gradations," makes Pater's almost ethereal and subjectivized world Darwin's as well. The language at certain points is surprisingly close, given the difference in the writers' ultimate aims and audience. The solidity and permanence of nouns are hostile to Pater's vision and enterprise, and recall Darwin's own words in the *Journal of Researches*: "Daily it is forced home on the mind of the geologist that nothing, not even the wind that blows, is so unstable as the level of the crust of this earth" (p. 323).

"The service of philosophy... towards the human spirit," claims Pater, "is to rouse, to startle it into sharp and eager observation."[23] (You will note that this passage is the source of the word "startle," which I have been sprinkling throughout this chapter.)[24] Darwin's job would seem to be more prosaic, but it was similarly to startle his readers into alternative possibilities, to alert them to the instability of the earth and of the organisms they have traditionally thought to be permanent.

Pater's famous insistence on catching each moment as it comes and burning with a hard gemlike flame suggests to me a charming but I think important anecdote from Janet Browne's biography of Darwin. Asked about his master's health, Darwin's gardener is said to have responded,

my poor master has been very sadly. I often wish he had something to do. He moons about in the garden, and I have seen him stand doing nothing before a flower for ten minutes at a time. If he only had something to do I really believe he would be better. (p. 460)[25]

The gardener cannot see what Darwin is doing, what "disciplined power of thought" went into Darwin's long observation of the flower, how what the gardener was watching was what George Eliot called in her description of Lydgate's scientifically important thoughts, "the last refinement of energy." It's hardly how the gardener would have put it, or Darwin himself, but there he was, standing unmoving before the flower, surely catching each moment as it came and "burning with a hard gemlike flame." Standing unmoving and "doing nothing," Darwin was piercing through the movements of time, seeing what others miss, recognizing barely visible relations, making the surface of the flower speak its secrets, the relations among its parts, its history, the soil in which it thrives, the insects to which it is mutually adapted, its means of reproduction, its relations within species, within genera, and with other species and genera. The intensity of Darwin's observations corresponds eerily to Pater's apparently aesthetically oriented language. "While all melts under our feet, we may well catch at any exquisite passion, or any contribution to knowledge that seems by a lifted horizon to set the spirit free for a moment" (*Renaissance*, 250). Little did Darwin's gardener know that Darwin's spirit was free for those moments, catching each moment and making his "contribution to knowledge."

As a scientist, Darwin was determined not to fall into what Pater would call our "failure," that is, our tendency to form habits, "for habit is relative to a stereotyped world, and meantime it is only the roughness of the eye that makes any two persons, things, situations, seem alike." What Darwin saw perhaps better than any scientist of his time was the uniqueness of every organism. His work was to catch at those strange variations, to develop an eye at least as acute as that of pigeon breeders, who could detect differences invisible to ordinary people, perhaps to develop an eye like that of brooding natural selection, who saw in the pigeon what even pigeon breeders could not see.

But, to return to the largest of the Darwinian paradoxes, the reading of Darwin that most upset habitual thinking was what Daniel Dennett now most gleefully asserts—that nature and all those wonderful processes are mindless. He notes Darwin's "shocking substitution of Absolute ignorance for Absolute Wisdom in the creation of the biosphere.... The very idea that all the works of human genius can be understood *in the end* to be mechanistically generated products of a cascade of generate-and-test algorithms arouses deep revulsion in many otherwise quite insightful, open-minded people." Revulsion from this vision was, indeed, the response of famous secular critics of Darwin, G. B. Shaw and Samuel Butler, but, while it may be a legitimate inference from his work, Darwin's writing rarely shaped the paradox as Dennett formulates it, and certainly not with Dennett's glee. As Richards has shown, Darwin almost never uses Dennett-style mechanical analogies, and at least in the first formulations of natural selection implied the possibility of progress and perfection. In the *Descent of Man* "progress" emerges unequivocally as the direction of the moral nature of humans, and yet this potential happy ending is indeed explained in entirely naturalistic terms—biological, not mechanical. So Darwin was part of the audience to be shocked by the bleak implications of his discoveries to which his "reason" was to take him.

But I invoke Dennett here once more because his deliberately provocative (and historically misleading) way of formulating Darwin's dangerous idea emphasizes the fundamental paradox at the heart of the vision that has been most widely diffused through Western culture. That paradox worked its way through an entire naturalist tradition of fiction, which in various ways dramatizes the incompatibility of consciousness and sensibility with the indifferent workings of nature. In the hands of some naturalist writers, and often of Thomas Hardy, the paradox became a tragic drama, a contest between human ideals and nature's indifference.

But here I want to return to the paradox as a form of the joke—the juxtaposition of improbables, the conjunction of contraries—mind and mindlessness. While for a writer like Hardy, the development of consciousness in humans is incompatible with adaptation to a meaningless world, and the drama of consciousness is almost always a disaster, for a writer like Pater, the drama of consciousness is virtually all there is of a life in which matter is so incompatible with mind that it cannot be known—it is only the thought of it that is known. The upshot of this is not inevitably tragic, but the creation of a new emphasis on consciousness as a creative force. Imagination presides over an otherwise unintelligible (or unattractive) world. What makes the world beautiful (as what made the tail of the Argus pheasant beautiful) is the mind, consciousness itself. It is that way of reacting to the paradox that I want to emphasize in considering, finally, the reach of Darwin's art towards the mind-oriented writers like Pater and most particularly (paradoxically?) Oscar Wilde. The greatest of paradoxes was mind itself, and it is to some of Darwin's writing about mind that I want now to turn.

Notes

1.   But to be fair, Dawkins does also want to insist on the exhilarating and enlivening aspects of science. He begins his book devoted to this view of things, in this way: "To accuse science of robbing life of the warmth that makes it worth living is so preposterously mistaken, so diametrically opposite to my own feelings and those of most working scientists, I am almost driven to the despair of which I am wrongly suspected. But in this book I shall try a more positive response, appealing to the sense of wonder in science because it is so sad to think what these complainers and naysayers are missing.... The feeling of awed wonder that science can give us is one of the highest experiences of which the human psyche is capable." What he objects to, and what spurs him to his emphases on the violence and destruction in the world is "saccharine false purposes," and "cosmic sentimentality." Richard Dawkins, *Unweaving the Rainbow: Science, Delusion, and the Appetite for Wonder* (Boston: Houghton Mifflin, 1998), pp. ix–x.

2. Richard Dawkins, *River Out of Eden* (New York: Basic Books, 1995), 96.

3. James Krasner argues that Darwin creates a point of view that is unlike the traditional scientist's (and novelist's) in that it is clearly limited: the world is too complex and entangled, the eye too imperfect a structure, to allow clear and complete vision. See *The Entangled Eye: Visual Perception and the Representation of Nature in Post-Darwinian Narrative* (Oxford: Oxford University Press, 1992).

4. A thorough discussion of this development would have to deal with Spencer's version of evolutionary theory, which was probably yet more influential culturally than Darwin's at the time. As Rick Rylance points out, "these two versions of the 'development hypothesis' existed side by side for at least a quarter of a century and their differences were frequently blurred even by informed commentators." Rick Rylance, *Victorian Psychology and British Culture, 1850-1880* (Oxford: Oxford University Press, 2000), 225. They were both, however, invested in a physical explanation of mind, and in the long run, Darwin's, as a careful examination of his language makes clear, was much more disturbing to conventional understandings of order than Spencer's, and despite Spencer's regular invocation of the Unknown, is much more dependent on a view that there is no mind in nature, except the mind as it develops along the chain of life from the "lowest" organism to the "human."

5. Gowan Dawson, *Darwin, Literature and Victorian Respectability* (Cambridge: Cambridge University Press, 2007).

6. While Darwin worked hard to dissociate himself from Lamarck's ideas, there is no doubt that he read him and something of the spirit of Lamarck's rejection of conventional biological thinking runs parallel to Darwin's own views. Identifying the various methods and uses of scientific knowledge, Lamarck distinguishes between "what belongs to artifice and what to nature." Startlingly, he insists that despite the usefulness of the current taxanomical orders, "Nothing of the kind is to be found in nature.... Nature has not really formed either classes, orders, families, genera or constant species but only individuals who succeed one another and resemble those from which they sprung." Although Lamarck sets out to find a "philosophical" way to describe the "natural series," and the "natural system," from the start he sets up a radical distrust of the system-making nature of the human mind, which is always working in its own "economic" self-interest and belies the overwhelming abundance and movement of the natural world. (J. B. Lamarck, *Zoological Philosophy: An Exposition with Regard to the Natural History of Animals* [Chicago: University of Chicago Press, 1984], see 20–2.)

7.   A fuller study of the relation of the "inward turn" of early modernist literature to the deep materialism of Darwin's way of looking at the world would require extensive treatment of the development of the modernist novel. That would require a book in itself. I am here, in a non-encyclopedic sort of way, trying only to suggest lines of influence that are very much worth following out in our efforts not only to understand Darwin's influence, but also the cultural implications of his work.

8.   Robert J. Richards, "Darwinian Enchantment," in *The Joy of Secularism*, ed. George Levine (Princeton: Princeton University Press, 2011), pp. 203–4.

9.   G. H. Lewes, *Studies in Animal Life* (London: Smith, Elder, and Co., 1862), 162.

10.   W. K. Clifford, "The Philosophy of the Pure Sciences," in *Lectures and Essays*, eds. Leslie Stephen and Sir Richard Pollock (London: MacMillan, 1901), v. I, 302. I do not want to claim that Darwin alone lies behind these arguments. Although Clifford was a Darwinian enthusiast, among the leaders of a Cambridge group affirming Darwinism, he was at least equally influenced by Helmholz and by Spencer's evolutionary theories. The point to be made here, however, is that the Darwinian ethos entailed a reading of the mind akin to Clifford's.

11.   Lewis Carroll, *Alice in Wonderland* ed. Donald J. Gray (New York; W. W Norton, 1975), 112.

12.   Smith's yet unpublished paper, from which this quotation is taken, "Darwin and the Sensation Novelists," makes an extremely important contribution to our understanding of Darwin's impact on the general culture, an impact that goes well beyond the scientific development of his evolutionary theory, and that extends beyond the "realist" novel, on which Gillian Beer and I have tended to focus. It is important to recognize here, and in other respects, how the very respectable Darwin provided crucial materials for much less respectable cultural phenomena.

13.   Charles Lyell, *Principles of Geology*, vol. 1 (Chicago: University of Chicago Press, [1830] 1990), 32.

14.   See Gillian Beer, "Darwin's Reading," in David Kohn (ed.), *The Darwinian Heritage* (Princeton: Princeton University Press, 1985), 54–7.

15.   In a chapter added to the third edition of *Darwin's Plots* (Cambridge: Cambridge University Press, 2009), Gillian Beer discusses Darwin's interest, from the start of his career, in the consciousness of animals. Humans are certainly not the only "conscious" creatures in the world, and Beer notes the variety of Darwin's "interest in sentient life forms and the peculiar forms of intimacy that this entails" (p. 248).

16. In response to my discussion of female choice in *Darwin Loves You*, Evelleen Richards has written to me a long, detailed, and extremely interesting critique, insisting that Darwin thought female choice unintelligent and capricious, and thought it destructive both among birds, who must then get so gaudy as to risk exposure to predators, and among humans. Her arguments are worthy of a long essay. I give here a small section of material that certainly will be soon published:

> I think the most distinctive aspect of your original argument is that with female choice Darwin made something rich and strange of his Victorian sexist assumptions, that it was the discrepancy between his conventional views and apparent instances of female agency in the development of male plumage and colour, which forced Darwin to the promotion and development of the notion of female choice. I am with you in this, though I disagree with your further claim that female choice somehow restored intention and intelligent direction to the theory of evolution. Further, while I would agree that Darwin's discussion of sexual selection was inflected by the language of Victorian courtship novels along with his notion of female beauty, my argument places much more significance on Darwin's lived encounter with the "ugly savage" and his early readings in ethnology and aesthetics (Darwin was familiar with some significant ones, such as Burke's *Inquiry*, Erasmus Darwin's *Zoonomia*, and Lawrence's "blasphemous" *Lectures*, even before he set foot on the *Beagle*). The notion of female coyness was crucial to sexual selection as Darwin understood it—without sexual coyness or a very Victorian modesty, females would mate indiscriminately with any or every male (just as males do with females), so there could be no niceties of choice or discrimination, especially not to the refined standards Darwin insisted upon. So coyness was vital and had to be retained.
>
> Female choice was not sexually active in Darwinian terms, but a more passive aesthetic exercise, and above all, capricious and non-functional. It might lead to beauty that could be appreciated by both birds and humans (as a beautiful dress might enhance the beauty of a woman), but it might equally lead to ugly wattles, clashing colours as in the Macaw, grotesque inflatable bladders, etc., or even be disabling as the peacock's tail. I agree that Darwin did regard the Argus pheasant plumage as beautiful, and I thought that your analysis of his prose showed that very well; but Darwin got a bit carried away with the Argus pheasant, or so his critics thought.

17. This citation is from the first edition of *The Descent of Man and Selection in Relation to Sex* (Princeton: Princeton University Press, [1871] 1981), Part II, ch. xiv, p. 142.

18. Ronald Thomas, "Wilkie Collins and the Sensation Novel," *in The Columbia History of the British Novel*, ed John Richeth (New York: Columbia University Press, 1994), 506.

19. Arthur Conan Doyle, *Through the Magic Door* (Pleasantville: Akadine Press, 1999; 1907), 245–4.

20. Arthur Conan Doyle, *Sherlock Holmes: The Complete Novels and Stories* (New York: Bantam Books, 1986), I, 14.

21. Walter Pater, *Plato and Platonism* (London: MacMillan and Co., 1907), 19–21.

22. Walter Pater, "Coleridge," *Appreciations: With an Essay on Tile* (London: MacMillan, 1922, 1889), 65.

23. Walter Pater, *The Renaissance: Studies in Art and Poetry* (London: MacMillan and Co., 1888, 1873), 249.

24. For an interesting discussion of the importance of what he calls for lack of a better word, "startlement," the condition of being startled, as a source of major value in a secular culture, see Paolo Costa, "Secular Wonder," in *The Joy of Secularism*, ed. George Levine (Princeton: Princeton University Press 2011).

25. Janet Browne, *Charles Darwin: The Power of Place* (New York: Alfred A. Knopf, 2002), 460.

5

Darwinian Mind and Wildean Paradox

I

The ultimate paradox of the Darwinian world is that mind, the product of mindlessness, becomes the greatest compensation for the loss of traditional notions of meaning and mindfulness, divinely inserted, in the world.[1] I have tried to indicate how the very "mindfulness" of Darwin's thought experiments through brilliant argumentation implicitly foregrounds the value of the mind in its efforts to find order in so multiple and various a world. While Darwin surely did not want to emphasize the degree to which his own mind in effect created the patterns he "observed" in nature, many of his readers—most particularly other writers—were likely to have recognized that. It is particularly interesting that when, at the end of the *Origin*, Darwin projects the consequences of his ideas into the "distant future," the first thing he says is that "Psychology will be based on a new foundation" (p. 488).

The expansion of physiological psychology in the last half of the nineteenth century, the new, not only Huxleyian emphasis on the "Physical Basis of Mind," the emergence of the journal *Mind*, and the inward turn in literature all not only corroborate Darwin's prediction but provide evidence of how important a job ideas of mind were doing in providing creative compensation for the disappearance of creative intelligence from behind the corporeal veil of this world. The great puzzle was how mind could emerge from mere matter, for it is the mind, and the mind of humans, that creates order and meaning. Darwin chose to begin where he usually did, not at the ultimate beginnings, but at the furthest point back in the history of life that could be accounted for without some ultimate theory of origins: "I have nothing to do with the origin of the primary mental powers, any more than I have with that of life" (p. 207). (It is no minor thing in one's quest for Darwinian paradoxes that *On the Origin of Species* has nothing to do with ultimate origins.)

Although Darwin did consider, particularly in his notebooks, philosophical issues, he was hardly a philosopher, except insofar as that which we call physical science was then still largely called "natural philosophy." He treads close to the large issues, and the question of mind/body, the question of how mind emerged from matter, could not have been uninteresting to him. But he steered away from the quagmire of deep philosophy. Among the complexities of developing Victorian preoccupation with mind there was certainly the influence of the philosophy of Kant, who thought that the human mind somehow matched the order of nature, although his ideas ramified out ambiguously in many even opposed directions.[2] It won't, moreover, seem a mere accident that Darwin had been a lover of Wordsworth's poetry, and would have known very well the idea that the mind half conceives and half creates the world. There certainly were Wordsworthian echoes in *The Journal of Researches*, but in Darwin's later work, even at its most literal and materialistic, that kind of idea is never overtly suggested. Rather, it is Darwin's writing itself, the way he proffers and contends with his persistently material world, that implies it. Through the great sweep of Victorian thought, the Romantic idea that mind alone could and did impose order on nature because it alone could create it was certainly given great force by the very emptying mind out of nature that marked the movement of Darwin's representation of the world.[3]

Before moving into the world of the most famous, mindful, and witty of paradox artists, Oscar Wilde, as an unlikely heir of Darwin's way of writing, and an unlikely participant in the work of compensation, I want to look at how Darwin creates and then treats that paradox of intelligent unintelligence. He does so in what should now be a very familiar way, confronting with awe, wonder, and surprise a seemingly inexplicable phenomenon, and then by way of evocation of the ordinary, explaining away the "wonder." The self-reflexiveness of evolutionary theory, the fact that talking about the universality of evolution inevitably entailed talking about humans themselves, was one of its most powerful and disturbing elements. Famously, Darwin preferred, in the *Origin*, to avoid direct confrontation with this particular disturbance, but everyone immediately understood it to be the most important issue. (The most famous anecdote about the initial hostility to and triumph of Darwin's *Origin*, was that of Bishop "Soapy Sam" Wilberforce, at a meeting in Oxford, asking Huxley on which side of his family he was descended from apes. Huxley is reported to have responded, "If there were an ancestor whom I should feel shame in recalling, it would rather be a man, a man of restless and versatile intellect, who, not content with…success in his own sphere of activity, plunges into scientific questions with which he has no real acquaintance, only to obscure them by an aimless rhetoric, and distract the attention of his hearers from the real points at issue by eloquent digressions and skilled appeals to religious prejudice.") Huxley may have won points on the Victorian scale of polite discourse and demonstration that evolutionists had higher moral standards than clergy, but the idea that Darwin's theory implied descent from ape-like creatures was immediately then and continues now to be the focus of the culture's view of Darwin.[4] Although it is true that Darwin cautiously avoided in the *Origin* every direct application to humans until he asserted at the very end that "Light will be thrown on man and his history" (p. 488), there are human analogies, with very human implications, in several passages before that ending. As Desmond and Moore insist, everyone knew that the true subject was always the origin of humans—and, in its distinctiveness, the origins of the human mind.

The mind may well have been Darwin's greatest "difficulty on theory." Shifting focus on Darwin's writing from the naturalistic ideas that provoked so much of the pessimistic literature at the end of the nineteenth century to the work of the mind as resistant to the material darkness is the most direct way to the Darwin I am trying to present in this book. We have seen something of such a shift already in Pater; it is also intimated, as I have suggested, in the intellectual pyrotechnics of the detective novel, which employs "scientific" techniques in demonstrating the power of intelligence to break through mysteries and put things to rights. The detective is the scientific artist of one of Victorian literature's most successful sub-genres. Darwin's apparently unartful and world-changing volumes offer an artful explanation of how the indifferent and trial-and-error world actually works, and they do it with a loving attention to particularities that emerges in a prose carefully honed and artful. The scenario, in various senses of the term, has "comic" implications.

The double movement of Darwin's prose is the stylistic enactment of the banishment of mind from nature, and at the same time, the dramatization of the mind's extraordinary power. It is also a method, or a succession of attitudes, at least, that seem designed to help put humans more at ease in a world so apparently incompatible with mind; a method that gives mind the power to put in order—or almost—a recalcitrant world ever entangling and complicating itself. The mind in Darwin's nature, of course, is not the divine mind but the creative human mind, and Darwin needed to know how it got there. It was an extraordinary development out of instinct and natural necessity, climbing up the organic ladder through gradations of human-like intellectual qualities (that Darwin takes pains to explain). He marvels at the instincts of ants and bees, and even wonders about the intelligence of worms, who contrive with apparent mindful intention to pull their leaves into the ground with the narrower end first. And in its fullest development, having emerged from such minimal flares of consciousness, mind somehow manages even to understand how it got to exist at all: through the workings of natural selection.

Forced to self-consciousness by the arguments themselves, Darwin's readers must watch with fascination the movements of Darwin's mind through language as he engages each problem. Of course, Darwin never points to his own virtuosity, but the experience of virtuosity recurs at almost every crux in his argument. We have seen it earlier in the passage quoted in chapter three where, defining natural selection for the first time, he remarkably turns the tables on his skeptics, putting the burden of proof on them and proving that it would be "extraordinary" if the agreed upon facts did *not* have the effects he describes. So it certainly is with his treatment of "instinct," the first step on the way to human intelligence, which requires a whole series of moves in order to establish in the very experience of the reader the idea that not only bodily characteristics, but mental ones can be inherited.

Darwin, of course had understood from the outset of his discovery of natural selection, that some day he would have to make the case for the evolution of mind.[5] So to arrive at the human mind, he begins in the *Origin* with instinct, a move that has the extra-scientific goal, as well, of softening up a readership that he knew would be reluctant to believe that not only bodily, but "spiritual" qualities could be inherited and governed by natural law. And he does so in the familiar pattern of the double movement: wonder–explanation–wonder. The wonderful (starting with the amazing instinct of the hive bee to produce the most perfect possible warehouse for his nectar) is transformed into a kind of accumulation of ordinariness, but an accumulation yet more astonishing than the original mystery.

"So wonderful an instinct as that of the hive-bee making its cells will probably have occurred to many readers as a difficulty sufficient to overthrow my whole theory" (p. 207). We can recognize the characteristic initiation of argument: there is our narrator, Darwin, confessing his problems and confronting the dangers by recognizing what others are likely to feel and believe. He knows that if he can't show that a quality of mind, however rudimentary, can be inherited and is a development through natural selection, his theory is shot long before he gets to the human. Like Wallace, he would have had to fall back on some extra-natural explanation, and the whole understructure of his actualist theory would have crumbled.

Darwin refuses to attempt a definition of instinct, and gropes at a general idea that it is "an action, which we ourselves should require experience to enable us to perform, when performed by an animal... without any experience, and when performed by many individuals in the same way, without their knowing for what purpose it is performed" (p. 207). It looks like consciousness but isn't—or is it? Here in instinct is the beginning of consciousness, an amazing phenomenon that is paradoxical in its very nature because it is an extant and recognizable example of mindless mind.

As we have seen, Darwin's prose almost always exploits the ubiquity of plurality, multiplicity, and blurred boundaries, and Darwin shows directly that instinct is not monolithic and inflexible as most people tend to think about it. No single definition will cover all instances. But Darwin also usually requires some confirming evidence and so he cites an expert, "Pierre Huber," in order to slip in a crucially important little fact: "A little dose... of judgment or reason, often comes into play, even in animals very low in the scale of nature." A careful reader here will prick up ears. Here is another Darwinian counter-intuitive idea—even the "lowest" animals may be said to be "reasonable," at least at times. One rather non-scientific response to this idea is to feel it as comic—a reasonable ant or rabbit or mouse! Gillian Beer singles out the example of Darwin's interest in sympathetic snails (Beer, *Darwin's Plots* 252; *Descent* I, 325). But, of course, while the attribution of intelligence to the lower animals is not intended to be comic in spirit it is a startling idea, startling in the way a good joke is, with the kind of effect that the talking dog jokes might evoke. This startling effect, however, is a consequence of Darwin's continuing full and respectful attention to the smallest of creatures (caterpillars figure importantly in the early parts of the discussion, for instance, but aphids will shortly turn up with rather Victorian manners, as well). It is difficult to forget his extraordinary attention on the *Beagle* to the behavior of spiders. So the argument itself and its startling implications emerge from a Darwinian habit of mind that is built into his prose at virtually every point.

Although Darwin was far more likely in the *Origin* to attend to ants, worms, and aphids than to humans, it is worth emphasizing yet again that implications for humans are not really hard to find. He notes at one point, "If a person be interrupted in a song, or in repeating anything by rote, he is generally forced to go back to recover the habitual train of thought" (p. 208). Nothing remarkable here, it would seem, except that such a comparison suggests that to understand instinct Darwin had already thought a lot about its functioning in humans. Of course, we know that he had been doing that, and much of the result of those observations emerges in the *Descent* and in *The Expression of the Emotions in Man and Animals*.[6] But such a simple-seeming analogy implicitly gives the lie to the idea that Darwin never talked about humans in the *Origin*. Each allusion has profound implications, for the implicit comparison is between a human and a non-human, so extravagantly different in some cases as the caterpillar. There is something in common between human habit and caterpillars building their "hammocks." The strategy of Darwin's argument, again, is to invoke shared experience to give experiential weight to the scientific arguments, and there is no avoiding human behavior as a point of reference.

Mozart makes his appearance shortly after the caterpillars: "If Mozart, instead of playing the pianoforte at three years old with wonderfully little practice, had played a tune with no practice at all, he might truly be said to have done so instinctively" (p. 209). This is an important part of Darwin's discussion of how instincts cannot be acquired by habit in one generation. Careful readers of the *Origin* will not have missed the way Darwin can draw on such human analogies to describe animal characteristics, and the affective and even moral implications of those analogies. Certainly, if our powers are continuous with those of caterpillars, and if it can be shown that caterpillars' or bees' instincts are the product of natural selection, it is not too long an inference to the idea that our most distinctive, apparently non-corporeal qualities, our ethics and our very art, are developments from natural selection.

To follow all the spins of analogy and all the details of evidence through Darwin's arguments here would be excessive. It should suffice to emphasize that he needs to show that "instincts are as important as corporeal structure for the welfare of each species, under its present condition of life," because if so, they are almost certainly going to be heritable and subject to natural selection; and—no matter how extravagant some of the comparisons may seem when taken from context—he does show it (p. 209). He needs also to show that instincts are governed by the same laws as bodily inheritances. They *are* inherited, they *do* vary. At this point, he can turn to some specific examples, but settles for only three—"the instinct which leads the cuckoo to lay her eggs in other birds' nests; the slave-making instincts of certain ants; and the comb-making power of the hive-bee." We can agree that there is nothing particularly poetic about this list, but at the same time it wonderfully suggests the fascination of nature's diversity, the potential significance of the smallest creatures, and an unpretentious but delicious engagement and fascination with natural phenomena. And the juxtaposition of such different details in nature suddenly makes connections among unlikely bedfellows—Darwin is always at work finding patterns in the entanglement, overwhelming as the entanglement will always continue to be. A sequence like this draws the reader deeply into the rich and various world of Darwinian nature.

With each of these three, the double movement comes into play. Darwin's treatment of the hive-bee, for example, begins in wonder: "He must be a dull man who can examine the exquisite structure of a [honey]comb, so beautifully adapted to its end, without

enthusiastic admiration" (p. 221). "Grant whatever instincts you
please, and it seems at first quite inconceivable how [bees] can
make all the necessary angles and planes, or even perceive when
they are correctly made." The difficulty seems overwhelming, but
not surprisingly, given our experience with Darwin, "the difficulty
is not as great as it at first appears." What follows, as Darwin
begins his quest for *verae causae*, is a consideration of a whole
range of extant bee species and their various devices for storing
honey, on the principle that natural selection always works by
"imperceptible gradations." The leap to the marvelous achievement of the hive-bees turns out not to be a leap but a walk up a
stepladder with rungs very close together. There are grades of
perfection of hives, in their honey-holding capacity. The explanation is long and detailed. In the fascinatingly precise and yet
speculative way Darwin so frequently uses, he also invokes geometry (with the advice of "Professor Miller of Cambridge") to try to
understand how the optimally efficient comb of hive-bees might
have developed from intentionless insects whose hives were far
from perfect. Such gradations exist in the varieties of bees now in
existence; they suggest the possibility of an historical movement up
the ladder of the degrees of perfection or imperfection we see in
bees now. So, as almost always, an analysis of relations across
extant organisms, this time bees, explains the historic process.
Darwin, like Holmes, moves from mystery, here of the hive-bees'
superb constructive powers, which would seem to imply some
miraculous intelligent intervention, to the much more unsurpris-
ing activities of other bees (context is here, as always in Darwin,
indispensable).

The intricate explanation takes the form of narrative once more.
Darwin shows how valuable it would be for bees, hard pressed for
nectar, to save energy and material and thus, "if a slight modification of her instinct led her to make her waxen cells near together so
as to intersect a little" it might make an enormous difference, a life-
and-death difference. Thereby the bees would save "some little
wax." Following this development in some detail, he concludes,

Thus, as I believe, the most wonderful of all known instincts, that of the hive-bee, can be explained by natural selection having taken advantage of numerous successive light modifications of simpler instincts; natural selection having by slow degrees, more and more perfectly, led the bees to sweep equal spheres at a given distance from each other in a double layer, and to build up and excavate the wax along the planes of intersection (p. 235).

All this happens although the bees are "no more knowing that they swept their spheres at one particular distance from each other, than they know what are the several angles of the hexagonal prisms and of the basal rhombic places" (p. 235). This touches on comedy, but it leaves the reader with a recognition that the bee is the perfect paradox—a conscious creature performing a brilliant act without knowing what it is doing, and all because of some small variation in inherited instinct.

Thinking of bees in geometry class evokes a smile; the incongruity, like the incongruities produced by almost all of Darwin's explanations, has the shape of a joke, though it was not perhaps intended to be funny. Surprise on one side followed by surprise on the other. The story is a good one, but by this time, after reading so many examples of this pattern of argument and accustoming ourselves to the quite beautiful precision of language, it is easy to forget that Darwin is, after all and once again, telling a story. The precision creates the illusion of direct empirical evidence, although it is always an imaginative leap, gathering information, putting it together analogically, ultimately making a grand, ordered imaginative conception. Darwin has provided a sensitive rendering of what the bees do, built into speculation about a history that can not be fully known—that is invoked, indeed, because Darwin refuses to accept the idea that the remarkable geometrical powers of the bees are part of the essential nature of the species, hive-bee. It had to be developed, hit or miss, through time. He gathers evidence enough, on the basis of his knowledge of the behavior of other bees, in particular, that the bees' geometric skills have not always been there. He thus writes a probable story constructed from his most careful attention to details—the shape and geometry of the hives, consultation with a mathematician, the behavior of other types of bees, the consideration of the expense of energy, and the possibilities of conservation. But like all of Darwin's explanations, materialist to the core, it also depends on the fundamental assumption of Lyellian actualism. He sees "real causes" now in operation and thus does not need to invoke an intelligent designer; telling a story, he shows that the extraordinary is also the product of the ordinary.

Nevertheless, the experience of the extraordinary remains, the shock of a world that can be seen as beautiful by human consciousness when the world itself is merely mindless (or instinctive) process. It is startling to realize that bees are dealing with "hexagonal prisms and...the basal rhombic places" and yet the knowledge of these things is not the bees', not even natural selection's, but rather Darwin's and his mathematician friend's. It is a thrilling story, and in fact a story of progress and possibility fulfilled. Thus, we arrive at my point: its shape is comic, like so much of the rest of the *Origin*. And comic in many ways. Bees as geometers.

Here, then, we might imagine that we are on the edge of *fin de siècle* irony, a comedy that juxtaposes the brilliance of the conception against the mere blindly natural world that is being conceived. Incongruity is normalized. Darwin's passage about the wonder of the bee stretches on for nine detailed and densely argued pages; as it withdraws intelligence from nature, it demonstrates it in the scientist/detective. It does not take much then to agree with Darwin's account of how nature might produce—does actually produce—extraordinarily beautiful and complex structures while remaining dumbly unaware that it is doing so. The counter-intuitive world becomes a reality under the pressure of a scrupulously careful describer and a wonderfully acrobatic thinker.

The final test of Darwin's stretch into the counter-intuitive comes in the last step in the Darwinian history of consciousness. Not only instinct but human consciousness itself grows from mindlessness. "If no organic being except man had possessed any mental power, or if his powers had been of a wholly different nature from those of the lower animals, then we should never have been able to convince ourselves that our high faculties had been gradually developed" (p. 86). We recognize here the synchronic strategy of other explanations—that is, while it is impossible really to know the history, we have evidence from the nature of other living organisms that the faculty, here consciousness, comes in many different grades, from simple instinct to complex instinct to habit to conscious choice. Darwin has shown that there are indeed many barely perceptible gradations of intelligence out there in the animal world. This is the groundwork for the demonstration that it all can happen without a teleology, without a designer.

But it is one thing to demonstrate that "man bears in his bodily structure clear traces of his descent from some lower form,"[7] quite another to say, with all the force of paradox, that "there is no fundamental difference between man and the higher mammals in their mental faculties" (p. 35). One can feel the abruptness and power of this assertion despite Darwin's very careful effort to build to it gradually. Although Darwin waited a long time to say it, and although the ground had been cleared for it by Huxley's *Man's Place in Nature* (1863), it was a difficult idea to reconcile to the culture's overwhelming sense of the enormous distance between human and animal intelligence. Darwin himself, beginning as he so often did, confronting an apparently insoluble difficulty, concedes that "man differs so greatly in his mental power from all other animals, there must be some error in this conclusion" (p. 34). Even Wallace, as we know, believed that there must, indeed, be an "error in this conclusion," and he challenges the very procedures of Darwin's argumentation. "Because man's physical structure has been developed from an animal form by natural selection, it does not necessarily follow that his mental nature, even though developed *pari passu* with it, has been developed by the same causes only."[8] Thus Darwin was contending not only with the inevitable resistance of his own culture, but with the objections of the very co-founder of the theory of "natural selection." If, in his discussion of eyes, or bees' hives, Darwin begins with marveling and wonder, what feelings might be appropriate to a contemplation of mind itself? Could mind be the product of not-mind? The formula wonder–explanation–wonder, the double movement, has a long way to go to work in this context.

We have seen that Darwin had, in the *Origin*, already gone a long way in his treatment of instinct, and the bees formed an important part of the argument. It was out of instinct, Darwin insisted, that human consciousness grew. Wallace, however, complains that Darwin simply does not produce the evidence to connect human consciousness with natural selection, and reading Darwin's analysis carefully, I find it hard to disagree with Wallace. Certainly there is no smoking gun—and yet Darwin, working as ever with elements of infinite complexity and the unknown, manages some extraordinary and very convincing intellectual acrobatics. In *Descent of Man* there is a long sequence of about twenty pages discussing animal intelligence that is free-wheeling, packed with different kinds of evidence, largely anecdotal, but certainly fascinating and engaging for any reader. Note how, moving beyond his point that it is sometimes difficult to distinguish the instinctive from actions driven by "free will" (recall how he has noted that there is a "dose of reason" in the instinctive acts of the most primitive of organisms), he offers this as one step in the argument:

The greater number of the more complex instincts appear to have been gained...through the natural selection of variations of simpler instinctive actions. Such variations appear to arise from the same unknown causes acting on the cerebral organization, which induce slight variations or individual differences in other parts of the body; and these variations, owing to our ignorance, are often said to arise spontaneously. We can, I think, come to no other conclusion with respect to the origin of the more complex instincts, when we reflect on the marvelous instincts of sterile worker-ants and bees, which leave no offspring to inherit the effects of experience and of modified habits. (p. 38)

As so often happens in his arguments, Darwin formulates from within his argument the kinds of conditions such arguments would have on a reader—like himself. That is, he claims that "we can, *I think*, come to no other conclusion." That curious element of personal presence is operating here, and it works to seduce the reader into feeling as Darwin himself feels about the evidence. There is no other way to take the evidence than Darwin's way. This strategy accompanies another familiar one, the invocation of "unknown" laws, not, of course, to disqualify or discredit the argument, but to increase its likelihood. "Little is known about the functions of the brain, but we can perceive that as the intellectual powers become highly developed, the various parts of the brain must be connected by the most intricate channels of intercommunication" (p. 38). All of this suggests, Darwin argues, that as the brain develops connections that would make each separate part "less well fitted to answer in a definite and uniform, that is, instinctive, manner to particular sensations or associations," we are on the way to full consciousness. The mind becomes more intelligent, less instinctive. It is a story full of "unknowns," quite elaborate and hardly proven. It is an imaginative story and a happy one.

Although the "unknown" lurks behind Darwin's storytelling, the success of his argument always depends on material for which he *can* find evidence, at least of the kind he provided for the development of the bees' hive-making precision. That is, he can show that, for example, "animals are excited by the same emotions as ourselves." In the *Descent*, Darwin showed himself particularly fond of anecdotes; stories operate everywhere, both as modes of explanation and as evidence for them. It is partly that characteristic that makes his work so interesting to non-scientists. Everybody loves a story, particularly Darwin. For the point that animals are excited by the same emotions that excite us, he entertainingly offers stories of dog's affection, their powers of sympathy and fidelity, their jealousy, and even their sense of humor. Beer points out that "the method of anecdote and observation, despite all its attendant difficulties of storymaking and anthropomorphism, gives presence to individual cases" (3rd edition, 254), and this, as I have been trying to show, intensifies the experiential quality of his arguments, persuading by narrative engagement as well as by logical argument.

This famous "scientific" text includes for evidence a delightful anecdote like this about the dog : "If a bit of stick or other such object be thrown to one, he will often carry it away for a short distance; and then squatting down with it on the ground close before him, will wait until his master comes quite close to take it away. The dog will then seize it and rush away in triumph, repeating the same manoeuvre, and evidently enjoying the practical joke" (p. 92). As elsewhere, Darwin comfortably takes the domestic and the ordinary to do heavy work on large issues, and smiles as he does so. He does not hesitate to use as evidence his own delighted experience with dogs.

The point is that in animals one can find all these human qualities: the capacity for "wonder," for "imitation," for "attention," memory, imagination, and even reason: "Animals may constantly be seen to pause, deliberate, and resolve. It is a significant fact, that the more the habits of any particular animal are studied by a naturalist, the more he attributes to reason and the less to unlearnt instincts" (p. 46). As we try to come to grips with Darwin's "art," and with the impact of that art on the culture at large, it becomes clear that in dealing with the absolutely most controversial element in his controversial theory, mind, and consciousness, he returns to storytelling—this time regarding the observed behavior, often received and narrated as anecdotes, of animals. Darwin loved dogs. He relies on material available to any non-scientific reader: the behavior of dogs and cats and animals in zoos and aquariums; and the strength of his argument depends on readers recognizing from their own experiences just the behavior Darwin describes, and being delighted by Darwin's sympathetic recording of it. Here, in a passage added in the second edition of the *Descent*, he even uses his own children as evidence. It is a very domestic science with which he builds the case for the natural history of the mind:

I kept a daily record of the actions of one of my infants, and when he was about eleven months old, and before he could speak a single word, I was continually struck with the greater quickness with which all sorts of objects and sounds were associated together in his mind, compared with that of the most intelligent dogs I ever knew. But the higher animals differ in exactly the same way in this power of association from those low in the scales, such as the pike, as well as in that of drawing inferences and of observation.[9]

There is virtually nothing in ordinary life that does not become part of this extended argument to demonstrate that indeed "there is no fundamental difference between man and the higher animals in their mental faculties." Son William Darwin, it turns out, is quicker of mind than a pike.

Surely, if we could step back from the solemn context of a scientific text and a culturally volatile subject (as many late Victorians were quite happy to do), we would recognize a delightful and comically surprising moment. While one might, like Wallace, have some severe doubts about the validity of the "evidence" produced here, it is hard not to enjoy in a quite simple way Darwin's rather bold application of his scientific and rhetorical method. He addresses with a perhaps unusual directness and confidence the conventional wisdom, those authors who "have insisted that man is divided by an insuperable barrier from all the lower animals in mental faculties," and he has "made a collection of about a score" of aphorisms to that end. "But they are almost worthless." Darwin's prose here makes the conventional wisdom almost laughable and it suggests something like Darwinian contempt as he provides a litany of the points that these aphorisms make:

It has been asserted that man alone is capable of progressive improvement; that he alone makes use of tools or fire, domesticates other animals, or possesses property; that no animal has the power of abstraction, or of forming general concepts, is self-conscious and comprehends itself; that no animal employs language; that man alone has a sense of beauty, is liable to caprice, has the feeling of gratitude, mystery, &c.; believes in God, or is endowed with a conscience. (p. 49)

Such popular misconceptions are still current, but against them Darwin offers twenty of his most delightful pages: in them we hear the storyteller and the lover of animals, with deep anthropomorphic sympathy for other beings, demonstrate how, despite the fact that in many of these things, humans certainly are far more developed, all of them can be found in animals. For the sake of identifying the tone, let one example suffice. Even on the matter of religion and belief in God, Darwin opens the possibilities for dogs. Here I indulge myself with a Darwinian moment I love:

The tendency in savages to imagine that natural objects and agencies are animated by spiritual or living essences, is perhaps illustrated by a little fact which I once noticed: my dog, a full-grown and very sensible animal, was lying on the lawn during a hot and still day; but at a little distance a slight breeze occasionally moved an open parasol, which would have been wholly disregarded by the dog had any one stood near it. As it was, every time that the parasol slightly moved, the dog growled fiercely and barked. He must, I think, have reasoned to himself in a rapid and unconscious manner that movement without any apparent cause indicated the presence of some strange living agent, and that no stranger had a right to be on his territory. (p. 67)

The almost throwaway, "sensible" dog (who wouldn't smile at this description?) on the verge of inventing a god of the parasol, indicates something of how Darwin normally looks at animals. He regards them with great sympathy, anthropomorphically, and attributes to them what we think of as distinctively human traits: "he must have reasoned to himself" may be unscientific but it is certainly delightful. Darwin cannot imagine behavior of this kind without imagining movements of reason like his own. It is anthropomorphic, but that after all is the point. In Darwin's writing, the experience of the animals is an experience of consanguinity with humans. Darwin will, of course, take the dog's somewhat

superstitious behavior in this instance through the small gradations, pointing out how "belief in spiritual agencies would easily
pass into the belief in the existence of one or more gods." So the
dog does good service in the demonstration of how, even in
matters of belief, animals share characteristics with humans.
(And I might suggest that it takes a particularly creative and
sympathetic mind to imagine a connection between the dog's
behavior and primitive religion.)

In the long hierarchy of consciousness from the lowest single-
celled organism up to "the most exalted object we are capable of
conceiving," Darwin finds continuity. He is perfectly happy to
share a history with his dog or with primates (although it must
be confessed he is not entirely happy to recognize—what he
must—that he shares a history too with the Fuegians).

He who has seen a savage in his native land will not feel much shame, if forced
to acknowledge that the blood of some more humble creature flows in his
veins. For my own part I would as soon be descended from that heroic little
monkey, who braved his dreaded enemy in order to save the life of his keeper,
or from that old baboon, who descending from the mountains carried away in
triumph his young comrade from a crowd of astonished dogs—as from a
savage who delights to torture his enemies, offers up bloody sacrifices,
practices infanticide without remorse, treats his wives like slaves, knows no
decency, and is haunted by the grossest superstitions.

Man may be excused for feeling some pride at having risen, though not
through his own exertions, to the very summit of the organic scale; and the
fact of his having thus risen, instead of having been aboriginally placed there,
may give him hopes for a still higher destiny in the distant future. (404–5)

There is no need to go on with these examples, or to pursue the
uncomfortable implications of Darwin's own preference of dogs to
Fuegians (that hurts Darwin lovers everywhere). But once again,
the Darwin normally perceived by our contemporary scientists as
disbelieving in progress is upbeat, looking toward a future perhaps
as far advanced over contemporary civilization as civilization is to

"savagery." The climax of Darwin's arguments and his experience of wonder and delight in the world points towards a far richer set of possibilities than is conventionally allowed. But even his revulsion from the Fuegians has some positive implications, for it also reflects his full awareness that they *are* humans too, consanguineous with himself and indeed related to him directly. We might lament Darwin's failure—perhaps because he knew he was too close to them—to extend to the Fuegians the sort of imaginative sympathy that marks his relationship to virtually every other living creature, but there is no denying the upward thrust of his prose, its celebratory engagement with the details of life.

Darwin's counter-intuitive list of qualities possessed by animals leads to the counter-intuitive generalization that just as there are gradations of sight that can take us from the blind mole to the sharp-eyed eagle, so we can move from instincts to mathematics (an absolute stumbling block for Wallace, by the way) to human intelligence, morality, and aesthetics. While I have myself here emphasized the delightful and perhaps unscientific qualities of Darwin's arguments about the sources of human intelligence, I want very strongly to agree with Robert J. Richards when he complains that "for the modern critic, it is easy to become bemused and then to cluck at Darwin's stories of the humanlike traits of animals, while missing the logical force that lies behind them."[10] The point here is not to worry out the validity of Darwin's ideas, since what I am most interested in is the texture of his prose, the quality of its affect, its experiential nature, the way in which the very means of argument was as important as his ideas to contemporary writers who actually read him and came to use his work. But fully to appreciate the nature of the prose, it is helpful to recognize, as Richards points out, that the anecdotal quality of many of the arguments was entirely up to Darwin's contemporary standards, since so much of it (we'll put aside the dog and the parasol for a moment) was derived from the best experts in the field. Moreover, it is important to note that Darwin's imputing ideas and feelings to animals was based on a principle still shared by modern scientists, that is, "we predicate psychological traits of other people, as well as of animals, on the basis of manifest behavioral responses and similarity of nervous systems" (p. 197).

As we reflect on the nature of Darwin's prose, it is worth taking seriously Richards' contention that on matters of religion, Darwin's strategy was "to corrode by insinuation and barely visible logic the usual barriers surrounding the discussion of religious belief" (p. 199). In this light, the dog and parasol sequence is worthy of Wilde himself, and perhaps funnier. Religion, on this account, is built out of doglike limitations as well as intuition. Among the imperceptible gradations that take us from instinct to religious belief, we have to pause at the level of Darwin's dog and realize that the dog's sense that something is moving the parasol and impinging on his territory is not disconnected from religious faith in higher gods!

II

The contrast between a mindless and indifferent world and human sensibility and intelligence can produce tragic forms, it is true, and the work of Thomas Hardy, which I will be looking at in another light in the next chapter, might be seen as the most overt engagement with this incongruity. As even the case of Hardy will show, it is equally likely to produce comic forms, emphasizing the incongruity and through recognition of that incongruity actually producing laughs, but much more frequently producing literature that tends to celebrate the creative power of human imagination and intelligence.

As I argued in the previous chapter, one quite reasonable move in response to the impermeability of the natural world to human control is to turn inward. That inward turn, so effectively anticipated and intimated by Pater's *Renaissance*, is already visible in George Eliot, whose external world still required full attention since, while it was extremely difficult to reconcile to human ideals, it seemed for George Eliot to embody a principle of "Nemesis."

That is to say, George Eliot had still not surrendered the possibility that moral order was built into the way the world works. However devious and circuitous its working, the world of George Eliot tends to respond in kind to the moral condition of her characters. But the tension in her works between the powers of character, mind, moral determination, and habit, and the constraining force of the outside world almost always dramatizes the triumph of that outer world while celebrating the richness, complexity, and potential moral rigor of the inner. The inner life of her characters is the true action of her narratives.

The mindlessness of a nature that produces a being capable of recognizing the mindlessness shifts the burden of value from God to man. It is there not only in George Eliot, but yet more intensely in Henry James, and even in the more action-filled narratives of Joseph Conrad, where the inner life remains the focus, although often impenetrable to others and even to the speculations of the narrators.

But let us turn now to the comic Darwin, the Darwin of paradox and delight in the incongruities, aberrations, and intricacies of the nature that mind reveals and labors to disentangle. Perhaps the best locus for the paradoxical turn of post-Darwinian literature is in Oscar Wilde's remarkable, comic, and theoretically impressive dialogue—"The Decay of Lying."

Darwin turned the world on its head and tried to stay quiet about it. Wilde spent all his time forcing people to notice its paradoxical nature and to laugh with it. Instead of the soul-raking "nature red in tooth and claw," he has his main speaker, Vivian, complain that "Nature is so uncomfortable. Grass is hard and lumpy and damp, and full of dreadful black insects."[11] The difference in tone does not belie the similarity in understanding of nature. If the incompatibility of nature with human satisfactions and expectations is the theme of naturalist literature, so is it with Wilde. Catastrophe in naturalist fiction transforms, in Wilde's deliberately snobbish and ironic dialogue, into the discomfort produced by "dreadful black insects." The tone, not the idea, is the difference, as is the attitude provoked by it. In both cases, the snob's and the tragedian's, nature is incompatible with human ideals of order and human aspirations to pleasure and satisfaction.

There are not many direct allusions to Darwin in Wilde's work, or even in his notes and letters, but there is no doubt that Darwin is a presence there. By the time Wilde came on the scene, Darwin was fully absorbed into British intellectual culture. It is interesting that for both Pater and Wilde, Darwin's ideas were partly assimilated to Hegelian and idealist philosophy, so that when one reads Pater one senses Darwin as a materialist who confirms the ideas of the great continental philosophers. Nevertheless, both Pater and Wilde were very interested in the work of modern science, and both greatly valued Darwin's scientific demonstration of a world in flux, a world of becoming.[12]

For Wilde, poetry and science grow from the same roots; the differentiation between them grows later. There can be little doubt that he read Darwin sympathetically, recognizing those visionary, imaginative qualities on which I have been focusing throughout this book. In "The Artist as Critic," Gilbert talks of how "it was reserved for a man of science to show us the supreme example of that 'sweet reasonableness' of which Arnold spoke so wisely, and alas! to so little effect. The author of the *Origin of Species* had, at any rate, the philosophic temper."[13] And although one is inclined to think of Wilde primarily as a celebrant of Art, with a capital A, his concerns for Art were always linked (if complexly) to ethics—the inutility of Art was a challenge to the values of a booming capitalist and bourgeois emphasis on use and money. On the same page in which he invokes a sweetly reasonable Darwin, Gilbert pontificates: "Ethics, like natural selection, make existence possible. Aesthetics, like sexual selection, make life lovely and wonderful, fill it with new forms, and give it progress, and variety and change." There can be little doubt that Wilde had read *Descent of Man*, and with an ear properly tuned to its comic celebration of diversity, growth, change, and the beauty of the natural world.

That Wilde's aesthetic was ethical in its energies and directed against the dominant social and economic forms of his time is clear from a whole range of arguments that Gilbert makes. At one point, complaining that "Facts are not merely finding a footing-place in history, but they are usurping the domain of Fancy," he laments the "vulgarizing" of mankind:

The crude commercialism of America, its materialising spirit, its indifference to the poetical side of things, and its lack of imagination and of high unattainable ideals, are entirely due to that country having adopted for its national hero a man who, according to his own confession, was incapable of telling a lie, and it is not too much to say that the story of George Washington and the cherry-tree has done more harm, and in a shorter space of time, than any other moral tale in the whole of literature. (p. 305)

Shocked at this gross exaggeration, Ernest exclaims, "My dear boy!" only to hear immediately the complicating point, that "the story of the cherry-tree is an absolute myth." Wilde doesn't make ethics easy. And while the whole manner of his writing is in polar opposition to the sober, steady production of "facts" that we find in Darwin, it takes only a moment's pause to recognize, first, that the effect toward which Wilde aims usually implies some impressively complicated idea that requires paradox for its expression, and second, that Darwin was *not* a Gradgrind grinding out facts but a brilliant rhetorician whose images of the natural world were almost entirely "invented," in the sense that they were works of the imagination, the product of thought experiments, hypotheses, analogy, metaphor, and, to be sure, keen observation. But remember that observation for Darwin always meant inventing hypotheses and asking questions.

At the end of "The Artist as Critic," Wilde makes clear that the intimations of Darwin's work, its tendency to celebrate the achievements of mind, to move the action inside, had critical importance for him. The paradoxical assertion of the primacy of criticism over art, which is a central motif of this dialogue, implies just this point about the primacy of consciousness. What makes the world stunning and marvelous is the human mind's capacity to see it that way, to think about it, to find order in it. And so Gilbert, who does seem to be Wilde's voice in this latter dialogue, says to Ernest:

One more thing I cannot help saying to you. You have spoken against Criticism as being a sterile thing. The nineteenth century is a turning point in history simply on account of the work of two men, Darwin and Renan, the one the critic of the Book of Nature, the other the critic of the books of God. Not to recognize this is to miss the meaning of one of the most important eras in the progress of the world. The Critical Spirit and the World-Spirit are one. (p. 407)

Thinking of Darwin as the "critic of the Book of Nature" is not such a bad idea. It is one step away from being, well, Oscar Wilde, whose critique of the book of nature runs through the other dialogue, "The Decay of Lying".

Wilde's speakers, elegant, snobbish, spoiled, seem hardly scientific, hardly the struggling protagonists of Gissing-like middle-class ordinariness, or the Hardyesque combatants against ill-fortune, or the tortured Conradian grotesques. While Darwin was overjoyed to have the opportunity to struggle through the wildernesses of South America, Gilbert is put off by the "dreadful black insects" in the grass. Nevertheless, what Darwin learned about "dreadful black insects" was very important to Wilde's way of looking at the world, and he builds his theory of art out of Darwinian materials.

Consider this: "My own experience is," says Vivian, "that the more we study Art, the less we care for nature. What Art really reveals to us is Nature's lack of design, her curious crudities, her extraordinary monotony, her absolutely unfinished condition." Darwin's science was doing very much the same thing, insisting particularly on nature's "unfinished condition." He recognized that things are only "perfect" in relation to their environments and that those environments change, that many parts of the organism, like eyes or wings or fingers, are jerry-built over millennia, transformed by the willy-nilly pressures of changes in climate or geological shifts. Hardy's Unfulfilled Intention looms over this drawing room scene, but Hardy's pessimistic sensibility can't get in. The infinitudes, unknowns, and entanglements that mark Darwin's nature are all there in Vivian's foppish complaint about nature's "absolutely unfinished condition." Against this mess of nature, Wilde builds up the fortress of Art, for Art, Vivian claims, "is our spirited protest, our gallant attempt to teach Nature her proper place" (p. 291). I would like to suggest that Darwin had already been doing that, though obviously in a different register. The *Origin* is a reading of nature that, first, registers its disorder and entangled abundance, and then tries to give it coherence and shape through the overriding theory of natural selection. Science, in that respect, and certainly Darwin's science, is a kind of Art.

Then, anticipating the argument I have developed out of Culler's point that the form of Darwin's writing is comic, Gilbert turns potential tragedy into comedy: "Nothing is more evident than that Nature hates Mind. Thinking is the most unhealthy thing in the world, and people die of it just as they die of any other disease. Fortunately, in England at any rate, thought is not catching" (p. 291).

This is irresistibly funny (at least to me), not simply in its delightful jab at "England," but in the way it takes a motif of the bleakest naturalism, that thinking is a kind of illness that makes plain our incompatibility with the natural world, and turns it into a kind of joke. It is a serious joke. (One is reminded here of the unselfconscious figures in some of Conrad's stories, figures like the captain in *Typhoon*, who emerge from crises undisturbed and successful just because they haven't the intelligence and sensitivity of the characters, like Lord Jim, in whom Conrad and his stories are really interested.) Nature hates mind. Vivian love Art, which he regularly defines as opposed to nature or, in "The Critic as Artist," as anticipating nature, in fact creating it. Both Darwin and Wordsworth lie behind this way of thinking. Art teaches nature its proper place. Hence, mind is the supreme value. And that surely is one of the major inferences one can make from a reading of Darwin's work, all the while that Darwin is trying energetically to keep that work objective, submissive to the facts. But we have seen how brilliantly Darwin learns to manipulate the facts, to string them together in ways that make his explanatory case about how the world really works. This is the work of the "critic" of nature.

Vivian's contempt for "realism," the tediousness of the everyday, is an aspect of his sense of the incompatibility of mind and art with a nature that is not designed. Things survive because of their usefulness and Wilde will have none of that. Ruskin, Jonathan Smith has reminded us, was deeply upset by Darwin's concentration on the sexuality of organisms; flowers, Ruskin insisted, are not for reproduction, but for the pleasure of human observers. Wilde takes the Ruskinian perspective, pushing it to places Ruskin wouldn't have wanted to go. So Vivian insists that

the only beautiful things . . . are the things that do not concern us. As long as a thing is useful or necessary to us, or affects us in any way, either for pain or for pleasure, or appeals strongly to our sympathies, or is a vital part of the environment in which we live, it is outside the proper sphere of art. (p. 299)

It is valuable in the light of Vivian's argument here to recall that when a characteristic of an organism is immediately useful to it, it is also outside the proper sphere of Darwin's critique. This is because Darwin sees that what is useful to an organism can tell us little about its history. It tells no story. But what remain as vestiges or useless parts reveal the true history—things that likely were useful to the organism's ancestors, things inherited but largely without function or even counter-productive. Darwin was interested in men's nipples, and humans' rudimentary tails, and woodpeckers that don't peck wood. Wilde's humor here parallels and seems in part a response to this paradoxical Darwinian fascination with what would seem useless.

Wilde's aestheticism affirms the Darwinian world in the act of rebelling against it. With God out of nature, and design intrinsic only to the human, Wilde affirms art as the antidote to the indifferent world, and replacement of the disappearing god. Even his deliberate and outrageous translation of Art into "lying" resonates in the context of evolutionary science, for recall that for Lamarck, science was a kind of convenient lie. He did not use the word, lie, of course, but he contrasted scientific ordering to the natural order, just as Wilde contrasts nature with lying. Wilde, of course, blithely turns away from "nature," for what he values is in the lie, which fills the mindless and indifferent world with meaning and value. It is, then, human inventiveness that does it.

The form in which this argument is made is the paradox, the inversion of normal ways of thinking: "One touch of Nature may make the whole world kin, but two touches of Nature will destroy any work of Art" (p. 301). Darwin's theory makes the whole world kin. Too much of Darwin's world, a material nature hostile to intelligence, will kill Art. It is not out of place here to think of the mindlessness that produces mind, or of the mind that can detect—or insert—meaning and history in the mindless movements of nature. Wordsworth, Wilde claims, "found in stones the sermons he had already hidden there" (p. 301). Meaning is in the mind that observes nature, not in nature itself. Art opposes itself to mindless nature.

So Vivan's nature is curiously like Darwin's. Those stones have meaning not only as Wordsworth imposes his ideas upon them but as Darwin works out their history and fills them with a past that tells us a lot about the world. But "Life, poor, probable, uninteresting human life—tired of repeating itself for the benefit of Mr. Herbert Spencer, scientific historians, and the compilers of statistics in general, will follow meekly after [the liar/artist] and try to produce, in her own simple and untutored way, some of the marvels of which he talks" (p. 305). And when Vivian speaks of the Greeks' relation to art, his ironies imply a whole ethic as well as aesthetic. "They knew that Life gains from Art not merely spirituality, depth, thought and feeling, soul-turmoil or soul-peace, but that she can form herself on the very lines and colours of art and can reproduce the dignity of Pheidias as well as the grace of Praxiteles" (p. 308).

As empiricist as Darwin, Wilde pushes to empiricism's paradoxical twin (or offspring), solipsism: experience registers reliably not the external world but one's own subjectivity; while we think we are observing objective reality we are, as Pater had it, registering our own impressions. Like Wordsworth, we are reading our own sermons in stones. Art is the construction of mind, a great lie, and it is Art that manages to make nature meaningful. Darwin, the scientist of hypotheses and thought experiments and anecdotes, is thus also an artist.

The form of paradox Wilde employs is more than a game. It suggests one way in which our culture has learned to handle the primal ignorance and indifference that Dennett discusses: Feuerbach was there before this, arguing passionately that Christianity and God are merely the projections of Man's best self into a natural world that contains no God. But the position follows from Darwin's rendering of nature, and we have seen how in *The Descent of Man* Darwin shows how consciousness, but also art and morality, grow naturally out of human instincts. For Darwin, morality can be detected in its early forms in the conflict of instincts: "a struggle may often be observed in animals between different instincts, or between an instinct and some habitual disposition; as when a dog rushes after a hare, is rebuked, pauses, hesitates, pursues again or returns ashamed to his master" (I, ch. iii, p. 83). Darwin also gives the example of the swallows and house-martins, which "frequently desert their tender young, leaving them to perish miserably in their nests" (p. 84). The "miserably," by the way is the vintage, sympathetic Darwin. The birds desert "their tender young" because they are pulled by an even stronger instinct, migration, which certainly is lifesaving for them. It is from such conflicting instincts, combined with even the most primitive capacity for memory, that, according to Darwin, conscience is produced (pp. 83–4).

However the story gets told, it is human consciousness that fills the world with value and meaning, and for Wilde, the instrument of value is consciousness itself. It may well be a development out of natural selection for utilitarian purposes, but it becomes human just when it transcends utility. Art is a manifestation of that consciousness, and a celebration of it.

"Where," Vivian asks, "if not from the Impressionists, do we get those wonderful brown fogs that come creeping down our streets, blurring the gas-lamps and changing the houses into monstrous shadows.... Things are because we see them, and what we see, and how we see it, depends on the Arts that have influenced us" (p. 312). In effect, Wilde is talking here, in the language of art, about what I have claimed Darwin did in the language of science. "To look at a thing," Vivan says, in a formulation that recalls both Ruskin's claims about seeing and Darwin's practice, "is very different from seeing a thing. One does not see anything until one sees its beauty." I hope I have demonstrated already that this is not an abandonment of Darwin but an extension of him: Darwin never sees anything unless he also hypothesizes and asks dozens of questions. It is true that Darwin would, to say the least, have been made uncomfortable by Vivian's exclamation, "things are because we see them, and what we see, and how we see it, depends on the Arts that have influenced us" (p. 312). But if we explore this quite serious extension of empiricism, I think we will find that it works with Darwin and his science a lot better than it first seems. The world, on Vivian's account, changes as its Art expands. If we take the leap, substitute the word "science" for "art," something quite striking happens: "when Art is more varied, Nature will, no doubt, be more varied." We have heard Darwin looking forward to future generations, who are not burdened by the limits of conventional perception, and who will thus be able to see the world differently. The world certainly did change under the pressure of Darwin's imagination of it, and then under the pressure of the science that followed upon his work.

In any case, it is not entirely coincidental that Darwin's theory of sexual selection is tied so closely to the beautiful, that is, to the human perception of what is extravagantly beautiful in nature, and the subsequent passion to understand where that beauty might have come from. Without Darwin's (and his culture's) appreciation of the apparently non-utilitarian loveliness of bird feathers, sexual selection would probably not have been thought of. Darwin's dogged pursuit of the meaning of his subjects implies intense engagement with them. Although the virtuosity of Darwin's investigations has not quite the flair of Wilde's, he too is a virtuoso of observation and imagination.

So, ironically, even as he fled the association, Darwin softened up the world for aestheticism and Wilde and he left us an art as dramatic and more breathtaking than Wilde's, less literally funny but structured like a titanic joke—a paradox. The titanic is the tiny; the tiny is the titanic. Life imitates Art.

Wilde's paradoxes play out some of the consequences of Darwin's writing in ways that point forward to the ironies and inwardness of modernist literature, to the kinds of narratives, the experiments with stream of consciousness, for example, that persistently make us aware that what is being seen and registered is filtered through a particular consciousness. Suzy Anger, who recently discussed this matter with me, has very helpfully pointed me toward a passage in Virginia Woolf's wonderful short story, "A Mark on the Wall," which reverberates with a heightened Darwinian sensibility and formulates a way of seeing strikingly similar to Wilde's:

It is curious how instinctively one protects the image of oneself from idolatry or any other handling that could make it ridiculous, or too unlike the original to be believed in any longer. Or is it not so very curious after all? It is a matter of great importance. Suppose the looking glass smashes, the image disappears, and the romantic figure with the green of forest depths all about it is there no longer, but only that shell of a person which is seen by other people—what an airless, shallow, bald, prominent world it becomes! A world not to be lived in.

As we face each other in omnibuses and underground railways we are looking into the mirror that accounts for the vagueness, the gleam of glassiness, in our eyes. And the novelists in future will realize more and more the importance of these reflections, for of course there is not one reflection but an almost infinite number; those are the depths they will explore, those the phantoms they will pursue, leaving the description of reality more and more out of their stories....[14]

Obviously, the texture of the prose is different, but the resistance to "reality," and all those black bugs that bother Wilde is here. That sense of Darwinian multiplicity is here also. But what matters is the art, the illusion, and although the paradox isn't directly formulated, in the end the "reality" becomes what art makes it. The tangled bank is part of our culture's imagination, though it may have started as only a metaphor for Darwin's vision of nature itself.

The passage I have just quoted comes in the midst of a story where Darwinian echoes are everywhere. We even have the cruder, more obvious reflection, "life being an affair of attack and defense after all" (p. 4). The mark on the wall evokes from Woolf a series of ruminations about the directionless changeability of the reality in which she sits. It is a story inside a particular mind at a particular moment, registering its own impermanence, seeking to grab onto anything (that mark on the wall, for instance), to hold still. It is a story that has picked up the disordered chancy nature of a nature (however beautiful it might be) that leaves traces, that turned Wilde to art—that turns Woolf to art:

To show how very little control of our possessions we have—what an accidental affair this living is after all our civilization—let me just count over a few of the things lost in one lifetime, beginning, for that seems always the most mysterious of losses—what cat would gnaw, what rat would nibble three pale blue canisters of book-binding tools? Then there were the bird cages, the iron hoops, the steel skates, the Queen Anne coal-scuttle, the bagatelle board, the hand organ—all gone, and jewels, too. Opals and emeralds, they lie about the roots of turnips. What a scraping paring affair it is to be sure! The wonder is that I've any clothes on my back, that I sit surrounded by solid furniture at this moment. Why, if one wants to compare life to anything, one must liken it to being blown through the Tube at fifty miles an hour—landing at the other end without a single hairpin in one's hair! Shot out at the feet of God entirely naked! Tumbling head over heels in the asphodel meadows like brown paper parcels pitched down a shoot in the post office! With one's hair flying back like the tail of a race-horse. Yes, that seems to express the rapidity of life, the perpetual waste and repair; all so casual, all so haphazard....

Of course, Woolf does very different things with this way of imagining than does Wilde. But the kinship is obvious. Nature is without form and direction, and yet it leaves traces and has a long history; the mind seeks direction, and for Wilde, of course, that direction is Art. And while Woolf's story seems half lamentation for the transience, for the accidental nature of things, it is, in the very fact that Woolf has turned it into a (very beautiful) short story a celebration of the tangled bank, after all. And the wonder of the story is just the fertility and creativeness of the writer's mind as it rummages in the largely forgotten, digging into the nitty gritty of the everyday, the cats' claws, the rats, the underground, the omnibuses—all the crude shell of reality that the narrator says is so bold and boring, too—and turning them into a stunning imagination of consciousness, a recognition, which is also formally beautiful, of that transience, and of the mind's fallibility. (That mark on the wall turns out after all to be a snail—how Darwinian. Too bad it wasn't a worm, but the point would be the same.)

If the story Darwin tells shapes the disorder, transience, impermanence, and entanglement into a scientifically coherent narrative, the "reality" that Woolf seeks to engage is the reality of the mind, the mind imagining and the mind not visible in the passenger across the aisle in the omnibus, only traceable in its material expression. One thinks here of Woolf's other wonderful essay, "Mr. Bennett and Mrs. Brown," where in effect she also criticizes attention to the mere surface of things, the "reality," the boring real that Wilde's persona despises, and seeks in Mrs Brown "the depths to explore," the reality of infinite possibilities that she sees to be the province of art. The kinship, without Wilde's ironies, of course, is striking; the connection between the mind-created world in its paradoxical forms that Wilde evokes and the modernist novels that Woolf was to write is unmistakable.

Wilde startles us with paradoxes (and in Woolf those paradoxes survive—the dull unreality of reality, the "airless, shallow, bald, prominent world," for example) that suddenly begin to make sense when seen in relation to each other, but that become more meaningful still if seen in the light of Darwin's transformation of the world and what might be called his comic vision—a vision that emerges from his keen and loving attention to nature and that runs against the conventions of seeing and meaning inspired by the dominant energies of the culture into which he was born. He is likely to have been alarmed by my alignment of him with Wilde. And since Woolf did not write novels with pretty women and happy endings, he probably wouldn't have cared much for her, either.

The paradox of his own position is that he was not only a child of the conventions of thinking and seeing that he challenged, but he was entirely devoted to *not* disrupting them socially, even as he created a literature that put enormous stress upon them. Darwin is accused of being father of a number of things beside his ten children—Social Darwinism, for example, evolutionary psychology, and ecology. Could I then add modernism? Wilde is certainly a true heir in that he was ready to *épater le bourgeois* and through a paradoxical comedy project a Darwinian vision on the conventions that he wanted very much to see rejected. Woolf's sense of transience and extraordinary attention to minutiae and pressure to make meaning emerge entirely from the mind—all this makes her too an heir to Darwin. The kinship, with all their differences, among Wilde, and Woolf, and Pater form a developing tradition whereby the action is moved inside, and the material reality (which paradoxically produces the "inside") is not what it seems—perhaps only a snail.

So Wilde was right to find in Darwin's prose startling paradoxes that could lead to keener recognition of the excitement and energy of the life that surges from the visual through imagination into language, and finds value in and care for this world of puzzling, lovely, flying feathers. It is perhaps the ultimate irony that the strategy of Wilde's work was to denigrate the nature that Darwin loved, but for Wilde what Darwin loved was the nature he brilliantly constructed out of the most imaginative and creative storytelling. And that story is not inevitably bleak.

The struggle with which Darwin is normally associated is just as often "mutual aid," and the order that emerges from such apparent randomness, perceived, perhaps created, like Wordsworth's stones, dazzles; the history of things pulsates through the minutest details of their appearance. Yes, many of the consequences of Darwinian complexity are not pleasant, but there is a comic story to be told from Darwin's way of seeing and arguing, one that Wilde figured out how to tell. And the world that Darwin bequeaths us, its ramifying branches and rastro drawing us back literally billions of years, is a gift that artists and writers continue to exploit and explore.

Notes

1.   In the essay added to *Darwin's Plots* in its third edition, Gillian Beer points out how for Darwin's loving disciple, George Romanes, concern with consciousness was connected to the "colossal losses entailed in the death of the individual" as an assuaging of the pain (p. 243), and focuses on the way Darwin (and Romanes) explored the presence of consciousness across the whole organic world (including, for Darwin, the sympathy even of snails), and the way by what I have been calling "imperceptible gradations" consciousness developed. The idea that consciousness compensates for the losses entailed in the world of Darwin's theory is the larger point of my explorations here.

2.   For an excellent consideration of the development of psychology among the Victorians, with considerable discussion of the role Darwin's theories played in the developments, see Rick Rylance, *Victorian Psychology and British Culture: 1850–1880* (Oxford: Oxford University Press, 2000). His

discussion of Kant's influence on developments in Victorian society are particularly to the point here. He talks briefly, for example, of Kant's avoidance of the "epistemological abyss his critical reasoning might open by insisting that innate categories of thought happily agree with the order of nature" (p. 50).

3. There is no space here, nor do I have the expertise to fill it, for a consideration of physiological psychology, even in the years preceding the publication of the *Origin*. There is already a large literature about it, and about Alexander Bain, the leading figure in the development of this form of psychology. One of the critical problems of physiological psychology was just, as Rick Rylance puts it, how to transfer "conceptual terms from accounts of the physical world to those of mind and consciousness" (p. 178). It is the mind/body problem now inserted into working psychology. One way to think of Darwin's arguments in the *Descent* is that they are attempts at such an explanation. It is important, however, to keep in mind that while I am in effect attributing the new emphasis on mind largely to Darwin, there was an important movement to explain consciousness in material terms that goes back to Hartley, James, and John Stuart Mill, and that developed significantly with Bain's thought. Bain's early important book, *The Senses and the Intellect* (1855), whose title clearly indicates the materialist basis of the psychology, was published four years before the *Origin*; and as Suzy Anger, who has been extremely helpful to me in discussion of this problem, points out, Bain had published a great deal before that date. Darwin's arguments certainly helped solidify the efforts to connect materialism with consciousness, but what matters for my arguments here is only that such materialist arguments threatened, as Huxley himself was very happy to concede, to produce deterministic explanations of what had been considered "free" will and thus to drain the world of value. Writers of all kinds were driven to recognize value and the possibilities of freedom, and could do so best by at once recognizing the power of the hard, unaccommodating actual, and celebrating the powers of consciousness, however paradoxically, to resist it.

4. Adrian Desmond and James Moore demonstrate clearly not only that everyone read the *Origin* as being primarily concerned with the human, but that Darwin knew that would be the case: "He knew everyone would read 'mankind' into the book. These twelve words meant that 'Man is in [the] same predicament with other animals.' This would be an open secret, as he admitted to Lyell." *Darwin's Sacred Cause: How a Hatred of Slavery Shaped Darwin's Views on Human Evolution* (Boston: Houghton Mifflin Harcourt, 2009), 310.

5. The strongest evidence for the centrality of questions of mind and behavior to Darwin's efforts to understand the origins of humanity is in the "M" and "N" notebooks, put together in those critical years in which his ideas about natural selection were being formed. They were begun on 15 July 1838. The materials from these notebooks went into the *Descent of Man*, although, of course, they would have been available to Darwin at the time of the writing of the *Origin*. See the introduction to these notebooks by Sandra Herbert and Paul Barrett, in *Charles Darwin's Notebooks: 1836-1844,* 517–19.

6. For evidence of the extraordinary delicacy of Darwin's psychological observations, the latter book would be extremely useful. I would just note here one passage that indicates that Darwin's sense of the subtlety of transitions and significance of minutiae was of enormous help to him in the reading of humans, as well as of non-humans. This passage shows some of the qualities of personal engagement and psychological acuity that one might seek in poetry and fiction:

> The study of expression is difficult, owing to the movements being often extremely slight, and of a fleeting nature. A difference may be clearly perceived, and yet it may be impossible, at least I have found it so, to state in what the difference consists. When we witness any deep emotion, our sympathy is so strongly excited, that close observation is forgotten or rendered almost impossible; of which fact I have had many curious proofs. Our imagination is another and still more serious source of error; for if from the nature of the cirumstances we expect to see any expression, we readily imagine its presence. (*The Expression of the Emotions in Man and Animals,* ed. Paul Ekman (Oxford: Oxford University Press, 1998), 19.

7. *The Descent of Man, and Selection in Relation to Sex,* 1st edn. 34.

8. A. R. Wallace, *Darwinism* (London: MacMillan, 1912, 1889), 463.

9. *The Descent of Man and Selection in Relation to Sex* (London: Penguin Books, 2004, 1879), 98–9. Darwin was one of the first scientists to study the behavior of infants, and kept a journal, eventually published as "Observations on Children," that is still studied and highly regarded by students of psychology and child development. The whole text of the "observations" is now available on line: RECORD: Darwin, C. R. [Notebook of observations on the Darwin children]. (1839–1856) CUL-DAR210.11.37 transcribed by Kees Rookmaaker. (Darwin Online, http://darwin-online.org.uk/)

10.   Robert J. Richards, *Darwin and the Emergence of Evolutionary Theories of Mind and Behavior* (Chicago: University of Chicago Press, 1989), 197.

11.   Oscar Wilde, "The Decay of Lying," *in Critical Writings of Oscar Wilde*, ed. Richard Ellman (New York: Random House, 1968), 291.

12.   For a useful discussion of the relation of Pater and Wilde's ideas to each other and to Darwin and idealism, see *Oscar Wilde's Oxford Notebooks*, eds. Philip E. Smith II and Michael Helfand (New York: Oxford University Press, 1989), esp. 43–5.

13.   Oscar Wilde, "The Artist as Critic," in *Critical Writings of Oscar Wilde*, 406. "The Artist as Critic" was published together with "The Decay of Lying" in a single volume, *Intentions*, in 1891.

14.   *The Mark on the Wall* (Richmond: Hogarth Press, 1919), 5.

6

Hardy's *Woodlanders* and the Darwinian Grotesque

While the tradition of paradox and irony, of the counterintuitive and the inward turn, continues deep into modernist literature, I want to conclude my discussion of the way the form of Darwin's writing helps shape cultural consciousness in the nineteenth century with a look at one novel by that unlikely suspect, Thomas Hardy. The Darwin whose paradoxical vision works its way into the aestheticism of Pater and Wilde appears once again in Hardy, and not only in the bleakly pessimistic version.

It has been easy enough to see Darwin (along with Greek tragedy and folk narratives, of course) in the famous last novels, *Tess of the D'Urbervilles* and *Jude the Obscure*. But it won't do to confine oneself to the inevitable catching of Darwinian strains in Hardy just where there is stress, competition, chance, struggle, and suffering. There is by now little need to elaborate or refine this side of the argument. We hear Darwin as the trees scrape against each other in the crowded forest of *The Woodlanders*; we hear Darwin in the squealing of the trapped rabbit in *Jude the Obscure*; we hear Darwin again and obviously in the embedded fossils into whose eyes Henry Knight stares as he clings for life to the side of the cliff, in *A Pair of Blue Eyes*; we hear him, alas, in virtually every phase of Jude's painful career and in his young son's suicide and murders.

But the Darwinian influence in Hardy, emerging not so much from his idea of evolution by natural selection as from the play of those ideas in the prose that created them, can produce rather different and more surprising effects. Gillian Beer captured some of what I am after in *Darwin's Plots*, in which she claims that

Alongside the emphasis on apprehension and anxiety, on inevitable over-throw long foreseen, persistingly evaded, there is, however, another prevailing sensation in Hardy's work equally strongly related to his understanding of Darwin. It is that of happiness. Alongside the doomed sense of weighted past and incipient conclusion, goes a sense of plenitude, and 'appetite for joy.' (p. 224)

That appetite for joy is certainly there, in *The Woodlanders* as well, and it emerges, among other things, from something like Darwinian loving attention to particulars, and natural observations of intense beauty. But beyond that, in place of the usual gloomy reading of Darwin's influence on Hardy's thought and art, I want to argue what I think is a more nuanced, more complex, and more interesting Hardy than that which usual reading allows. *The Woodlanders* is a book that resists the relatively easy placing of Hardy's novels as "tragic." It is, when looked at carefully, and particularly when looked at through Darwinian eyes, a very strange and unpredictable book, usually seen as just on the margins of the great Victorian canon of novels. Despite an ending that cannot be simply categorized as tragic, nor yet comic, it may be the most Darwinian of all.

Of course, everyone who talks about it has to talk about Darwinian images here and there, and I'll point to some of them. Hardy was, after all, consistently troubled by the inward turn, as I have been calling it, but also a participant in it. Consciousness, he seemed to believe, was no blessing to the human, but its antithesis.

Describing Clym Yeobright late in *The Return of the Native,* Hardy writes that Clym "already showed that thought is a disease of the flesh, and indirectly bore evidence that ideal physical beauty is incompatible with emotional development and a full recognition of the coil of things."[1]

The mind, which opened the spaces to art in Wilde's thinking, conflicts with the body in Hardy's. While these are, in fact, two sides of the same condition, the effects are obviously very different. But despite the fact that Hardy's novels do seem to take the paradox very hard, he and Wilde were not that far apart; the play of mind and art as much as the indifference (or beauty) of nature were of central concern to both of them, and even in his dark narratives there is something quite different from the determinist tragedy as Tess surrenders to the President of the Immortals, or Henchard goes off to die alone, or Jude dies unaided in the hearing of cheering students.

Hardy, however, always resisted criticism of his works that focused on their philosophy or ideas. Consider this famous passage from Hardy's notebooks:

As in looking at a carpet, by following one colour a certain pattern is suggested, by following another colour, another; so in life the seer should watch that pattern among general things which his idiosyncrasy moves him to observe, and describe that alone. This is, quite accurately, a going to Nature; yet the result is no mere photography, but purely the product of the writer's own mind. (LWTH, p. 158)[2]

The parallel with Vivian's talk in "The Decay of Lying" is striking, but Hardy avoids Wildean paradox, conceding that there is a "going to Nature," and yet insisting that art is "the product of the writer's own mind." Implicitly, the nature the artist might be said to represent is then also a "product of the writer's own mind." There is a similar parallel with Pater: to see the thing as in itself it seems to me to be—these are Hardy's seemings.

In Pater, recall, the "Conclusion" to the *Renaissance* comes almost as a moral injunction. One's responsibility is to catch each moment as it comes, as Darwin catches his moments watching flowers before the uncomprehending gardener's eyes. Like Darwin, Hardy claims to be "going to nature," yet the order of nature, following Darwin, easily becomes a creation of the mind since the real thing, if we can so call it, has neither rhyme nor reason, being only a tangled bank that requires either Darwin or Hardy to interpret and render beautiful. As Perry Meisel put it in his excellent study of Hardy (he too noting a Paterian connection), "headquarters was slowly seen to shift to the inner life of the individual",[3] and the Darwinian Hardy joins that movement in late Victorian literature that was throwing the drama inside.

The tragic Hardy clearly has a feel for the comic Darwin's counter-intuitive vision. The pressures of nature against the ideals of the mind increase the intensity of the work of the mind. While Hardy is normally not regarded as a novelist of the inner life, like George Eliot, say, or Henry James, his books are a constant struggle between inner and outer and, probably more than either Eliot or James, his focus was on the deep internalized ideals of his central character. One can see it even in tragic narratives, as we watch the world through Angel Clare's eyes when he discovers that his "Tess" has revealed herself not to be his ideal but a woman of flesh and blood with a past:

the complexion even of external things seemed to suffer transmutation as her announcement progressed. The fire in the grate looked impish, demoniacally funny, as if it did not care in the least about her strait. The fender grinned idly, as if it too, did not care. The light from the water-bottle was merely engaged in a chromatic problem. All material objects around announced their irresponsibility with terrible iteration.[4]

The objects of daily life that exist here are only fragments of Angel's consciousness, for the action is entirely inside. More striking still in emphasizing the degree to which Hardy is preoccupied with mind and ideals is Angel's response to Tess: "You were one person: now you are another" (p. 226). And Tess's helpless response: "I thought, Angel, that you loved me—me, my very self." The "very self" is accessible, if at all, only to the self. For Angel, Tess's reality was the idea of Tess: "the woman I have been loving is not you."

While this notorious sequence in *Tess* might seem simply to reflect one character's failure, it is central to Hardy's work, which is, indeed, as he insisted, a set of "seemings." This is most obvious in Hardy's even stranger last novel, *The Well-Beloved*, which plays out Pierston's imagination of the ideal women in a set of three narratives, in which, one by one, generation by generation, they fail to live up to that ideal. What Clym experiences as a tragic tension between the corporeal and the ideal, *The Well-Beloved* plays out in an ambivalent tension between reality and ideal, leaving Pierston a more or less forgotten artist who at the end engages in some modernizing and improvement of old pipes in town and old houses.

The Woodlanders touches only slightly on this intense inwardness that creates and then is disenchanted by reality. Nevertheless, in the light of this aspect of Hardy's quite Wildean way of seeing, Marty South's dogged devotion to Giles Winterbourne becomes something rather different from the touching heroic fidelity it feels at first to be. Nevertheless, *The Woodlanders* provides strong evidence of the various ways that Darwin's vision, beyond the "idea" of natural selection, infected the way later writers would imagine their world. The most important Darwinian elements in that book are in its very form and in the curious way the "inside" action works while the narrative is so brilliantly and movingly attentive to the natural world. It is a book that also resists the clarity of tragedy. If any Victorian novel can suggest how odd Darwin's vision was for the Victorians, how subversive of the structures of their thought and their imagination of story, and how close he brought them to a very different modernity, I think *The Woodlanders* is it. It is a book laden with modernist ironies, drifting from image to image, with a lyrical seriousness that makes it at times meltingly beautiful, and that justifies the *almost* tragic conclusion, in which we hear Marty South's extraordinary elegy for Giles Winterborne, and her determination not to forget him as all others do. But these images constitute only one part—if the most powerful and memorable—of the book. While at no point does Hardy lose his virtually instinctive commitment to attend to the minutiae of nature, with an attentiveness akin to Darwin's own, the book notoriously slides away quite frequently into social comedy, into sexual play, into farce, into melodrama, and into scenes that might be thought invented to parallel the low life moments of comic relief that mark Shakespeare's comedies. A drunken and bruised Dr Fitzpiers is "flung" from the mare he is sharing, unbeknownst to him, with his father-in-law, Melbury, having lamented aloud that he is stuck with his wife, Grace Melbury, but would rather be with the superior Mrs Charmond. This is a moment from eighteenth-century farce (p. 256). Or again, the jealous Grace notices that Suke Damson's teeth are excellent and easily infers from their health that her husband has been casually unfaithful; Grace and Mrs Charmond stumble into each other in the woods and warm each other into life as Grace learns that Fitzpiers has "had" the older woman, whose hair was really Marty South's. Or how to reconcile these with the moment in which the three women unselfconsciously and touchingly cluster around the bed of the possibly dying philosophical roué, in a moment that has the shape of restoration comedy, unsentimentally representing a fundamental natural sexual energy, but has its pathos as well. There are pastoral scenes, and there is the melodramatic figure, onstage for one puzzling moment, who will eventually, offstage, kill Mrs Charmond. There is the curious long episode in which Giles dies outside his home while Grace protects her respectability within. (This seems to belong to that tendency to impute the ideal to a figure who, like Tess, in all corporeal solidity, resists it.) And there are frustrated romantic scenes between Giles and Grace, which are a mark of virtually all Hardy novels.

How disturbing *Jude the Obscure* was to Victorian respectability is well known. But while contemporaries by and large admired *The Woodlanders*, many rightly found it shoc*king as well. In some ways, I would argue, The Woodlanders is more subversive than Jude* of Victorian assumptions about narrative, which were linked, of course, to Victorian assumptions about virtue and value. R. H. Hutton had it about right:

This is a very powerful book, and as disagreeable as it is powerful. It is a picture of shameless falsehood, levity, and infidelity, followed by no true repentance, and yet crowned at the end with perfect success; nor does Mr. Hardy seem to paint his picture in any spirit of indignation that redeems the moral drift of the book.[5]

Early critics also register what they take to be the difference in quality between the representation of the woodland characters and that of the unpleasant modern ones, most particularly Fitzpiers and Mrs Charmond, who, as Coventry Patmore wrote, "give an ill-flavour to the whole book."[6]

The Woodlanders is not an in-your-face attack on Victorian morality, as, to many, *Jude* seemed to have been. But for critics as sensitive and intelligent as Hutton, it was an alarmingly disruptive book, "powerful . . . and as disagreeable as it is powerful." What makes its subversiveness most visible is its ending, which utterly denies the expectations of both the characters and the reader. Surely, it came as a shock to Hutton and many Victorian readers. My guess is that even now, for most readers trained in the ways of Victorian fiction, it comes as a surprise. We know Hardy well enough to know that a happy ending is out of the question, and yet the last episodes edge toward farce and conclude with recon-ciliation of the sort that one might have expected from Comedy. Grace returns to the scapegrace Fitzpiers, she doesn't get killed by the mantrap, she suffers very little guilt for the death of Giles, and Grace's father and the locals predict a henpecked Fitzpiers and squabbly marriage. Such an ending is consonant with what might be thought of as most subversive about Darwin's thought—not its tragic implications, but its potential for moral indifference, its refusal of poetic justice, its narrative that crowns as supreme value not morality but survival. There is no indignation that redeems the moral drift of the book.

There is a famous passage that can suggest some of the ways in which a fully Darwinian mode of thought might make sense of the generic chaos of the book and mark it off from the dominant conventions of the Victorian novel. Right at the start, the narrator reminds us, as Giles and Marty talk:

Hardly anything could be more isolated or more self-contained than the lives of these two walking here in the lonely hour before day, when grey shades, material and mental, are so very gray. And yet, looked at in a certain way, their lonely courses formed no detached design at all, but were part of the pattern in the great web of human doings then weaving in both hemispheres, from the White Sea to Cape Horn.[7]

This echoes the romantic and highly moralized organicism of, say, Carlyle's *Sartor Resartus*, or of virtually any high Victorian novel, like *Bleak House* or *Middlemarch*. But it is an echo as well of that undemonstrative celebration of mind that we have seen emerging from Darwinian writing. The narrator makes the connections that none of its actors can. Without pushing the analogy too far, I would suggest that this kind of commentary in Hardy suggests just that pattern of mindlessness producing mind that Darwin's work describes. As the bees unwittingly create their hives, so the characters take part in "no detached design," but one that encompasses the whole world. Everything, the argument goes, is connected to everything else (that Rube Goldberg sort of world), and there are deep moral consequences to this cosmic fact.

But it is not George Eliot's all-knowing angel Uriel, looking down with supernatural vision on the complex interweaving of human life and seeing meaning where humans can see none; it is a detached, Hardyesque narrator, more focused on esthetic pattern than on moral significance. Connectedness in *The Woodlanders* has narrative but not moral consequences. In George Eliot and Dickens, connections with unknown others enforces fundamental Christian notions of responsibility for one's own actions and responsibility to others, but in *The Woodlanders*, the strange diversity of action and character seems more gratuitous the more tight the connections are. What in George Eliot might be described as Nemesis, in *The Woodlanders* is felt like unhappy coincidence. It is the chance whose working Darwin cannot explain.

Connections are there regardless of the interests and knowledge of the characters, in an impersonal and unintentional, quite Darwinian, way. Only the "artist" perceives the design in events and movements that must feel and be mere "Hap" to the people who are seen as part of that design. For the characters, the gratuitousness of these inescapable connections inspires fear at being caught out at something the community might disapprove. They are likely, as is Mrs Charmond, to be much more worried about the respectability of their actions than about their morality. Will they be *seen* breaking the canons of community respectability. The answer is usually yes.

The Woodlanders' strange conjunction of Hardyesque motifs: the splendid and memorable detailed images of the natural world; the material traces of the past that are visible on fence posts and in their materiality register a form of consciousness, or memory; the pains in the aching bones of Mr Melbury; the imposition of the ideal on a universally material world; and finally the extraordinary casual juxtaposition of ostensibly incompatible genres—all of these gather additional significance if they are seen in the light of the pervasive presence of Darwin in Hardy's imagination.

Of course, influence is a complicated thing, and it's possible that Hardy would have written as he did without Darwin. Michael Millgate has observed, for example, that Hardy "found little difficulty in ranging ideas newly derived from Darwin and Huxley alongside the necessitarian views already instilled in him by both the peasant fatalism of his upbringing and the tragic patterns of the Greek dramatists."[8]

But it's not Greek tragedy we find in *The Woodlanders*, or not much. It's true that as early as the end of the first chapter, the narrator makes a move already familiar in the novel tradition, suggesting that the quotidian places and details of ordinary life are as latent with epic and tragic significance as the more heroic details of other genres. This is a characteristic typical of the English novel tradition, from Fielding to Hardy; it is an aspect of its realism; and, as we have seen, it is a central method and technique of Darwin himself, who, like Hardy, takes seriously ants, aphids, and worms, just as Hardy attends to "ephemera" and slugs. The narrator notes how "from time to time, no less than in other places, dramas of grandeur and unity truly Sophoclean are enacted in the real, by virtue of the concentrated passions and closely-knit interdependence of the lives therein" (p. 8). Yet, if the book closes with a Sophoclean Marty South, the central narrative ends in the comedy of an at least momentarily henpecked Fitzpiers, with Grace "queening it" so as to "freeze yer blood" (p. 365).

All of these extraordinary and unpredictable conjunctions are evidence for Culler's case that the form of Darwin's writing was "comic," and that his influence on later writing was in the direction of comedy. Darwinian reversals are everywhere; the book is full of Darwinian attention to nature's details; and of inversions of the idea of "design." The nature of *The Woodlanders* refuses to respond to human need, plays tricks to curdle the hair of a natural theologian, and is full of singular oddities. Hardy sounds rather like the Darwin who, claims Paul Barrett, "was always looking out for natural phenomena that would be imperfect or pointless from the point of view of an all-knowing Designer."[9] Darwin's mixed world, in which categories normally viewed as distinct are intimately connected, is reflected not only in the natural world represented in Hardy's novel, but in the very unstable form of it, as genres seem to clash from page to page.

And here too is the kind of Darwinian grotesque about which Jonathan Smith has written. Conventional proprieties about the body are threatened in Hardy's world, and expectations of unified identity, consistent behavior and feeling, are constantly disappointed. Much of Darwin's work, claims Smith, "can be characterized as grotesque." And he lists, along with the fascination with worm castings and their work, "the bizarre sexual arrangements of barnacles and orchids; the outré forms of fancy pigeons; the extravagant plumage, ornament, and weaponry of male birds; the hideous facial expression [in illustrations for *The Expression of Emotions*] of Duchenne's galvanized old man; the elaborate traps of insectiveorous plants" (*Charles Darwin and Victorian Visual Culture* p. 249).

Darwin's interest in the grotesque starts early enough, in *The Voyage of the Beagle*, as he discusses the emergence of life on a recent volcanic island, where he concludes: "I fear it destroys the poetry of the story to find, that these little vile insects [e.g. dung-covered wood lice and spiders], should thus take possession before the cocoa-nut tree and other noble plants have appeared." (p. 11) Darwin's prose is full of this sort of bathos throughout his career. A writer saturated with Darwin's thought and ways of arguing would have recognized how central the grotesque is to his world's development, and what emerged in Wilde as satire or parody, emerges in the great reversals that mark most of Hardy's fiction. Human *idea* and aspirations do not match material reality. This can turn out either tragic or funny, or both.

I want now to look in some detail at *The Woodlanders*, from this Darwinian perspective, and see if in that light its generic confusions can make sense. There are, of course, Darwinian themes, too, as in the famous passage in which Hardy detects

the Unfulfilled Intention, which makes life what it is . . . as obvious as it could be among the depraved crowds of a city slum. The leaf was deformed, the curve was crippled, the taper was interrupted; the lichen ate the vigour of the stalk, and the ivy slowly strangled to death the promising sapling. (p. 52)

Clearly Darwinian as this is, it is important to note the formal twist by which Hardy replicates the tendency of Darwin's prose to leap taxonomic boundaries. The woods are *like* the city. Hardy is not merely looking at the woods but seeing them against an ideal—the very strategy of satire and irony. The pastoral woods become an urban slum and the Wordsworthian romantic ideal slides into the method of late nineteenth-century urban realism.[10]

The most morally disturbing element of *The Woodlanders* is not the affirmation of the Darwinian struggle for existence, but rather, the juxtaposition of supposedly distinct categories; in Darwin there is the implicit juxtaposition, without moral judgment, of human and animal (spelled out at last in *The Descent of Man*); in *The Woodlanders* there is the juxtaposition of a traditionally virtuous and innocent protagonist (Giles) with a traditionally wicked rake (Fitzpiers)—and with no moral judgment. *The Woodlanders* is Darwinian in form and mode because it enacts in startling juxtapositions comic grotesqueries, pastoral ideals, and low comedy, with a good dose of country tragic and the detailed vision of poetry.

The novel's first two chapters brilliantly open the world of contradictory genres and disturbing juxtapositions, most forcefully because that world is so precisely described and so imaginatively embodied in the deeply pastoral place, Little Hintock, Hardy's non-urban settings obviously allow for much fuller looks at nature, but they also guarantee that the inevitable crossings among members of a very small population will have terrible consequences. The opening is a quietly spectacular tour de force of Hardyesque perceptions, locating Little Hintock in the deep isolation of the woods, bringing into it a stranger utterly incapable of finding his bearings without help, suggesting in virtually every line the mysteries of perception and the pervasiveness of human vulnerability in nature. The narrator's focus on perceiving and being perceived registers with a virtually trembling sensibility the way no world is self-contained, every world and every act inside that world is likely to be perceived—and intruded upon—by another. When the barber Percomb enters Marty's cottage while, unawares, she works at cutting spars, she exclaims, "Oh, Mr. Percomb, how you frightened me!" His quick reply reverberates throughout the novel: "You should shut your door—then you'd hear folk open it" (p. 11). Every world is vulnerable to intrusion from others.

Like the juxtapositions in the earliest scene, the other generic disturbances in *The Woodlanders* seem, for the most part, very "natural."[11] Part of the power of the novel (and then its ultimately disturbing ramifications) is that it is easy not to pause over the differences, but simply to take them as narrative necessities. Hardy incorporates the Darwinian grotesque vision into the natural flow of his narrative, which is so richly and precisely filled with the texture of the woodlands that it creates a feeling of authenticity that is not disrupted by the various generic dances the characters living inside that world perform. The grotesque and the comic, the blurring of distinction between human and animal is in the book's texture. As Darwin eradicated the divide between the human and the animal (and beyond that, between the animal and the vegetable), for Hardy, humans are not more important than the bugs, horses, and birds that seem constant companions to the humans, whether the humans are always aware of them or not. Human failure to note them is often thick with moral implications. And it need not be an easy moral matter, for those who do not notice are as often the protagonists as the villains. In *Tess*, the "pure woman," moves along the grass in gauzy skirts like those of the other dairy maids, as "the innumerable flies and butterflies" are brushed up by the "gauzy skirts" of the dairy maids; and "unable to escape," the insects "remained caged in the transparent tissues as in an aviary."[12]

Odd and grotesque juxtapositions are everywhere in *The Woodlanders*. Its ultimate form, if one had to find a category, would have to be comic, both in the sense that it concludes with something that (in other circumstances, at least) would be taken as a happy ending, the reuniting of lovers, and in the sense that it is filled with grotesque details, some of which are recognizably comic—and funny—although in Hardy, laughter is almost always associated also with sadness and loss. It is not only the usual country bump-kin choral comedy, to which Hardy frequently resorts, but it is associated with virtually every class represented in the book. It's there in the terrified Grammer Oliver, worrying about the brain she has promised to Fitzpiers, in the awkward Giles preparing the house for Grace and Mr Melbury's visit, in the absurdly philosoph-ical Fitzpiers speculating in bad German philosophy while longing to be back in civilization, in the pathetic Mrs Charmond, pursued by the melodramatic gentleman from South Carolina.

The pastoral hero, Giles, is as vulnerable to grotesque comedy as those who work for and with him. The party he gives to impress Grace is full of pathos and moments from stage comedy: the chairs are over-oiled so that Grace's dress is stained (it's "not a new one" she reassures the mortified Giles); Creedle splashes Grace's face with grease from the dish he brings in on a three-legged pot; Giles was in a "half-unconscious state" and "did not know that he was eating mouthfuls of bread and nothing else." Throughout, the other guests talk in a dialect with a vulgarity that Mr Melbury takes as offensive to his newly returned daughter, and as beneath her (pp. 73–4).

With Fitzpiers the parody takes us into the heart of Hardy's own deepest concerns, for Fitzpiers, one recalls, enters the novel as a young man like George Eliot's Lydgate speculating on the relations between mind and body, unlike Lydgate positing an unbridgeable gulf. But the comic, satiric grotesquerie of the situation emerges most fully when Grammer Oliver tries to explain to Grace what Fitzpiers has said to her: The unlikelihood that Grammer Oliver, realistically speaking, could drop from her own dialect and be able to quote Fitzpiers' pretentious narcissistic philosophy is itself an indication of the post-Darwinian extravagance of the reversals here. But she does quote him as telling her: "Ah Grammer...Let me tell you that Everything is Nothing. There's only Me and Not Me in the whole world" (p. 49). All of this watered down German romantic philosophy, coming from the mouth of a lazy and hedonistic child of the aristocracy and addressed to a superstitious old woman whose brain he offers to buy, is clearly a cause for laughter. If Grammer Oliver takes it all seriously, Hardy's sense here of the absurdity of the contrasts and contradictions is primary—it is very clearly a clash of distinct cultures turned into something like burlesque.

The further irony, however, is that much that Fitzpiers is given to say in his isolation is serious enough and relevant enough to the enterprise of the book to be worth attention, even (and particularly) in the light of Hardy's determination to parody it. Fitzpiers, the philosopher, is most particularly a sexual being. Grace is overwhelmed sexually by his mere presence (poor Giles!—the echoes here of Bathsheba, Gabriel, and Sergeant Troy are unmistakable, and Sergeant Troy was no philosopher). Fitzpiers is immediately attracted to the "bouncing" young Suke Damson (p. 111) and doesn't hesitate to roll in the hay with her. The core of sensuality that marks his being is ironically played against his philosophical aspirations, and Hardy notes how, as he is about to go on his rounds, he puts aside a book by a "German metaphysician" that he had been reading: "for the doctor was not a practical man, except by fits, and much preferred the ideal world to the real, and the discovery of principles to their application" (p. 112).

But Fitzpiers is clearly right about himself, for example, when he puts it abstractly to precisely the wrong person (Giles), that isolated people

get charged with emotive fluid like a Leyden jar with electric. Human love is a subjective thing.... it is joy accompanied by an idea which we project against any suitable object in the line of our vision ... I am in love with something in my own head and no thing-in-itself outside it. (p. 115)

We can recognize here the same kinds of solipsistic thinking we have already seen in Pater's "Conclusion" to the *Renaissance*. Reality is inside the mind, not outside it. As we have seen, Angel Clare too was in love with something in his own head, to Tess-in-herself outside it. Not to press the philosophy too far, it is still clear that Fitzpiers is thinking like the German philosopher, Feuerbach, who argued that humans project their own ideal on the "outside" world, and then worship it—or in Fitzpiers' case, make love to it. In *The Woodlanders*, the absurdity emerges not from the ideas but from the way they are formulated, the context in which they emerge, and the narrative ease with which philosophy here can be read as lazy self-justification. The grotesque comedy emerges from the form more than from the substance of the ideas. Hardy's own skepticism about philosophical thought is enacted by embodiment, by recognizing the context in which the ideas are articulated, and the range of corporeal matters that tincture the ideas. Narrative allows Hardy just that ironic detachment that turn his serious ideas into comedy, and subjects them to the hard force of the material world, whose pervasive secular presence is part of the endowment of post-Darwinian consciousness.

The comedy is extended in Fitzpiers' actual courtship of Grace. When Grace comes to ask him to release Grammer Oliver from her promise, Fitzpiers is quick to seduce her with his ideas. He leads her to his microscope, in which she sees some "cellular tissue" that turns out to be from John South's brain. Grace's shock is the occasion for another bit of philosophy: "Here am I," he said, "endeavouring to carry on simultaneously the study of physiology and transcendental philosophy, the material world and ideal, so as to discover if possible a point of contact between them; and your finer sense is offended" (p. 133). It is absurd *and* right, an echo of things Hardy had worked through in many novels. And it is difficult not to think of this as something of a parody of George Eliot writing in *Middlemarch* about Dr Lydgate, who, in his search for the primitive tissue, seeks also to "pierce the obscurity of those minute processes which prepare human misery and joy, those invisible thoroughfares that are the first lurking-places of anguish, mania, and crime, that delicate poise and transition which determine the growth of happy or unhappy consciousness."[13]

But the parody in *The Woodlanders* is of ideas that are central to it. The post-Darwinian imagination of the narrator (as well as of Fitzpiers, one can presume) is imbued with a sense that all of life can be traced out in the material world. That is the fundamental "trope" of Darwinian argument. How then to account for, how to deal with, consciousness, art, love, morality? *The Woodlanders* in its narrative way is thick with the ironies we have seen played out comically in Wilde, as they both try to come to terms with the incompatibility between consciousness and matter, between the social and the natural.

The body affirms itself everywhere, and as it does (as with the fragment of John South's brain) the grotesque is at work virtually everywhere; human and tree, human and pigeon, human and mere matter are confused. The woods themselves are inhuman, not the Wordsworthian pastoral space for contemplation and calm. The contrast between material and ideal is present in even the smallest of details. The narrator even notes, for example, that the "green shades" of the trees in spring "disagreed with the complexion of the girls who walked there" (p. 144). But there are many more extended moments, as, for example, the scene in which Grace returns to the woods looking for her purse and Fitzpiers is sitting dreamily abstracted from the physical context, contemplating settling down in the "quiet domesticity" of this rural world. The awkward movements of incipient lovers begins, interrupted by this detail: "A diversion was created by the accident of two birds, that had either been roosting above their heads or nesting there, tumbling over one another into the hot ashes at their feet, apparently engrossed in a desperate quarrel that prevented the use of their wings" (p. 143).

This absurd juxtaposition of bird and human is preceded by an even more exaggerated moment. There are moments when the woods emerge not in pastoral calm but as a place where violence and grotesquerie dominate. Take this fine description of the "barking season":

Each tree doomed to the flaying process was first attacked by Upjohn. With a small bill-hook he carefully freed the collar of the tree from twigs and patches of moss which encrusted it to a height of a foot or two above the ground, an operation comparable to the "little Toilette" of the executioner's victim. After this it was barked in its erect position to a point as high as a man could reach. If a fine product of vegetable nature could ever be said to look ridiculous it was the case now, when the oak stood naked-legged, as if ashamed, till the axe-man came and cut a ring around it, and the two Timothys finished the work with the cross-cut saw. As soon as it had fallen the barkers attacked it like locusts.... (p. 136)

The analogy between swarming humans and swarming insects is not accidental. The metaphorical language is clear that this is a ruthlessly cruel (and, yes, Darwinian) activity. This is how the woodspeople make their living: they flay, they execute, and they descend like locusts. And the trees are given a kind of conscious-ness for a moment to extend the violence and the comedy—they are "ashamed" of their stripped condition (a line that comments more than indirectly on the anomaly of human consciousness and aspirations to respectability). The passage may be quaint but it is also grotesque—the juxtapositions are extravagant and at the same time comic.

But Hardy was himself uneasy about the way he ended the book. Writing to various of his friends and acquaintances, he warns: "It is rather a failure at the end."[14] As the characters inside the narrative are surprised by Grace's return to Fitzpiers, so, as we've seen, Hardy's first audience tended to be positively offended. It may be that Hardy believed that the whole narrative, from Giles's death, to the working of the mantrap (yet another bit of melodrama turned to comedy), to Grace's renewed acceptance of Fitzpiers, failed to live up to the idea of the book with which he had begun. He may have been made uneasy by the mixture of modes, which came so naturally to him. John Bayley suggests that Hardy's criticism of the book's ending was defensive and disingenuous, and that its refusal (Grace's refusal in particular) of a more straightforwardly daring and potentially tragic ending was the rejection of an ideal incon-sistent with the generic mix of the novel (pp. 193–4). It is charac-teristic of Hardy as writer and as narrator that he deeply admires the heroic (and inevitably catastrophic[15]) actions of his protagonists, but that he fears such actions and represents figures—like Farfrae, in *The Mayor of Casterbridge*—who find ways to survive. Survival always entails compromise, and a character incapable of excess, or heroic aspiration. Grace's nature, as she is imagined in the novel, is fully consistent with the novel's generic mixture, and makes the heroic—except in the passive form—an impossibility. The ending is the only right one in a novel characterized by such mixed modes.

Heroes are likely to have a hard time surviving and multiplying in Darwin's world, where survival and reproduction constitute success. But there is virtually no reproductive success in *The Woodlanders*. Or rather, it would be better to say, the protagonists have virtually no reproductive success. Although the traditional reading of the book emphasizes its preoccupation with the death of the ancient woodland culture before the thrust of modernity, it is important to note that even Grace and Fitzpiers, and certainly Mrs Charmond, produce no offspring, and it is they who are at least symbolically responsible for the decline of woodland culture. The woods, it would seem, continue to propagate, but for a novel so richly engaged in the life of the woods, it offers a peculiarly barren landscape. One notes the strange failure of the garden plots of Little Hintock, plots that "were planted year after year with that curious mechanical regularity of country people in the face of hopelessness," but "no vegetables would grow for the dripping" of the trees (p. 123). The small dying area in remote England is losing out in the battle for existence.

Recall that in his essay, "On Some of the Conditions of Mental Development," Clifford begins and ends with two statements that reflect the patterns of reversals and shiftings that I have been discussing. The first sentence is: "If you carefully consider what you have done most often during this day, I think you can hardly avoid being drawn to this conclusion: that you really have done nothing else from morning to night but *change your mind*."[16] It is easy to recognize here the sort of reversal that Culler takes as distinctive of late-century Darwinian form. Playing on the idiom, "change your mind," making it quite literal, Clifford produces yet another paradox. But it is a serious play on words, for the very condition of consciousness is changing one's mind. And at the end of *The Woodlanders*, that paradox, in both form and substance, seems to be instantiated. It suggests dramatically that the stability we take as normal—our sense of the coherence of our selves, the continuity of our being—is self-evidently false. The world of consciousness, at least, is made up of just the sorts of paradoxical or apparently contradictory movements that I have been suggesting characterize the mixed generic mode that we find in *The Woodlanders*.

There seems to be even more of the Cliffordian point of view in the novel's ending, as well. Here, again, is Clifford's conclusion to the essay:

If we consider that a race, in proportion as it is plastic and capable of change, may be regarded as young and vigorous, while a race that is fixed, persistent in form, unable to change, is as surely effete, worn out, in peril of extinction; we shall see, I think, the immense importance to a nation of checking the growth of conventionalities. It is quite possible for conventional rules of action and conventional habits of thought to get such power that progress is impossible, and the nation only fit to be improved away. In the face of such a danger *it is not right to be proper.* (p. 117—italics in original)

Despite a very important tonal difference and a perhaps too eager effort to shock, the passage has an unmistakable application to *The Woodlanders*. I am not trying to suggest that Hardy was following Clifford directly, but we have already seen how Clifford speaks with a kind of Darwinian voice (of course, he was equally influenced by Herbert Spencer). Clifford and Hardy seem to share the post-Darwinian sense that convention (or stasis) is utterly inadequate to the complexities of living and, indeed, to survival. They share as well (though in quite different literary registers) the desire to turn conventional wisdom about the world upside down.

Clifford was busy being an iconoclast; Hardy was yet more Darwinian because he worked very hard to maintain his respectability and not to sound like one. Thus, Clifford plays out his paradoxes loudly, insisting that precisely the sort of behavior that led Little Hintock gardeners to keep planting in impossible spots and that led Giles in effect to kill himself by disguising from Grace his illness and discomfort and keeping her respectable, and that leads Marty South to her dogged loyalty to Giles, will shortly wipe out the whole woodland culture. Reading the novel from this perspective, one would find that the emotional and moral force of the endings become exactly the reverse of what they seem directed to be—that is, moral and heroic. As an artist, Hardy seems morally ambivalent about the whole set of actions. But within a Darwinian scheme of plasticity as against conventionality, it is even possible that the moral value of the constantly adapting and mind-changing Fitzpiers/Grace world is greater than the moral value of the woodland protagonists, who cling to their ideals—and to their conventions. Virtue would seem to be on the side of the dogged and persistent, but survival is on the side of transient visitors to the woodlands. So, on the one hand, Giles sacrifices his life for the woman he loves (but rather for a social ideal that she herself will betray), and Marty's elegy is by far the most moving and powerful moment in the book; on the other hand, Fitzpiers, the adulterer, and Grace, the indecisive, succeed.

Within the Clifford scheme (a reasonable inference from the Darwinian one), virtue lies with the amoral plasticity that allows change. And it is difficult in any reading of the novel not to feel something other than virtuous heroism in Marty's barren loyalty. The very power of her speech lies in its latent paradox, and there is even something of Browning's Porphyria's Lover in Marty's elegy: "'Now, my own love,' she whispered, 'you are mine, and on'y mine for she has forgot 'ee at last, although for her you died'" (p. 367). Marty, whose only sexual characteristic is her luxuriant hair, will die barren. Her lyrical invocation of her own and Giles' remarkable power to plant things and to make them grow is both moving and an idealization of Giles, a projection of her frustrated desire on a man who, as he moves through the novel, is often bumbling and mistaken, and ideal only at the moment when he sacrifices himself for Grace. Marty's love is for a dead man, for only when he is dead can she possess him.

Against Marty's marvelous absence of plasticity, Grace's capacity to "change her mind" is played out in a scene out of another sort of romance. There is the comic/melodramatic setting of the mantrap for Fitzpiers (and the introduction of a characteristic Hardyesque *chance* that draws Grace toward the machine), but most of this is played out offstage. What we see onstage is the reaffirmation of the powerful sexual attraction between Fitzpiers and Grace (the reverse of the totally asexual relation between Giles and Marty and the virtually asexual relation—with one illegitimate kiss—between Giles and Grace). The reconciliation takes place outside of consciousness, choice, or even respectability. When Fitzpiers discovers that Grace has not been caught in the trap, he springs to his feet "and his next act was no less unpremeditated by him than it was irresistible by her, and would have been so by any woman not of Amazonian strength. He clasped his arms completely round, pressed her to his breast, and kissed her passionately" (p. 356).

Grace and Fitzpiers, unselfconsciously united again, find themselves once more in a green shade. Hardy's instinct for the image that speaks emerges again as the two "noticed that they were in an encircled glade in the densest part of the wood," at a moment in that "transient period" of May—"an exceptionally balmy evening"—"when beech trees have suddenly unfolded large limp young leaves of the softness of butterflies wings" and the boughs hung low so "that it was as if they were in a great green vase" (p. 358). The rich sensuousness and fertility of that green moment also suggest something about the sensuous and impermanent relation to come between Fitzpiers and Grace, something about how tightly the relationship is tied to material rather than conventionally romantic or moral conditions.

Although the novel, then moves to a close with a formal reversal of generic expectation and with the comic and cynical interpretations of the woodland chorus, it ends with Marty's elegy. That juxtaposition further emphasizes the generic instability of the book. Grace and Fitzpiers do escape the rigidities of Little Hintock and move, with a plasticity about which, surely, Hardy was ambivalent, into a modernity that bears the lines of life after all. The reversals are complete, and Grace's father, deeply disillusioned by Grace's choice, sullenly understates the novel's and the reader's understanding of the reuniting of Grace and Fitzpiers. "I have been misled in this," he says, as he hastens away from the daughter in whom he had invested the deepest feelings of his life.

Darwin lurks behind all of these elements of the novel: the radical materialism implicit in the Darwinian way of viewing the world; the reversals and boundary blurrings that mark his representation of nature; the emphasis on the corporeal—as in the unequivocal sexual power that pulls Grace and Fitzpiers together; the frustration of intention and consciousness; the disparity between human conceptions and material reality.

Hardy, like Darwin, preferred not to violate the conventions although he knew, with Darwin, that only violation of those conventions made life possible. The remarriage of Grace and Fitzpiers is evidence of plasticity and thus a means of survival, even if it violates the moral norms. It is the ultimate joke of a novel that refuses the tragedy that the "Unfulfilled Intention" would seem elsewhere to have demanded. Hardy took the mindless fusions of Darwinian processes not only to register the grinding competitiveness and tragic fatalism of all forms of life, but to set in motion a narrative that refuses to stand still for genre, that breaks the boundaries between tragic and comic, farce and melodrama, and that repudiates the tragic even as it enacts it. Darwin, who to Hardy and other writers, seemed the very source of tragic vision, lies behind this subversion of Victorian narrative form, and behind its ironies, its paradoxes and its turn to what we can now recognize as modernism. Emptying nature of the Romantic value that the

Victorian novel had virtually always sought in it, *The Woodlanders* dramatizes the inward turn that locates value, as Wilde and Pater did, in the workings of human consciousness. Life itself proceeds, violating conventions as it goes, with amoral energy and power.

Notes

1.   Thomas Hardy, *Return of the Native*, (Harmondsworth: Penguin Books, 1982), part 2, ch. 6.

2.   Michael Millgate (ed.), The Life and Work of Thomas Hardy, (Athens; GA: University of Georgia Press, 1985), 158.

3.   Perry Meisel, Thomas Hardy. The Return of the Repressed, A Study of the Major Fiction (New Haven: Yale University Press, 1972), 159.

4.   *Tess of the D'Urbervilles*, (Oxford: Oxford University Press, 1988), Part 5, ch 35, p. 225.

5.   R. H. Hutton, review from the *Spectator*, in R. G. Cox (ed.), *Thomas Hardy: The Critical Heritage* (London: Routledge & Kegan Paul, 1970), 142.

6.   R. G. Cox (ed.), *Thomas Hardy. The Critical Heritage* (London: Routledge & Kegan Paul, 1970) 142.

7.   Thomas Hardy, *The Woodlanders* (London: Penguin Books, 1998), 22.

8.   Michael Millgate, *Thomas Hardy: His Career as a Novelist* (New York: Raxtom House, 1971), 32.

9.   Charles Darwin, *Metaphysics, Materialism, and the Evolution of Mind*, transcribed and annotated by Paul H. Barrett (Chicago: University of Chicago Press, 1974), 66.

10.   Discussing a Hardy poem that works out some of the details that were so significant in *The Woodlanders*, Roger Ebbatson notes what would be an appropriate description of the novel: "The poem rejects what Hardy took to be the Wordsworthian position and states his own Darwinism, with its insistence upon combat as a valid metaphor for life. The poem concludes: "But having entered in,| Great growths and small|Show them to men akin—| Combatants all!|Sycamore shoulders oak,|Bines the slim sapling yoke,|Ivy-spun halters choke|Elms stout and tall." See Roger Ebbatson, *The Evolutionary Self: Hardy, Forster, Lawrence* (Brighton: Harvester Press, 1982), 5.

11. One would have to except the forced narrative of the stranger from South Carolina, an element so oddly intruded into the text that it always remains offstage. John Bayley points out that "Hardy himself seems neither to know nor to care that comic, pastoral, pathetic and tragic modes—to name only the most obvious ones—are all collectively at work." John Bayley, "A Social Comedy? On re-reading *The Woodlanders*," reprinted in Dale Kramer, *Critical Essays on Thomas Hardy: The Novels* (Boston: G. K. Hall, 1990), 191.

12. Thomas Hardy, *Tess of the Durbervilles* (London: Penguin Books, 1983), p. 146.

13. George Eliot, *Middlemarch: A Study of Provincial Life* (Oxford: Oxford University Press, 1996), 162.

14. Thomas Hardy, *Thomas Hardy: Selected Letters*, ed. Michael Millgate (Oxford: Clarendon Press, 1990), 54.

15. See *The Realistic Imagination* (Chicago: University of Chicago Press, 1981) for a discussion of Hardy's relation to the heroic ideal.

16. W. K. Clifford, *Lectures and Essays* (London: MacMillan and Co., 1901), vol. 1, 79.

Coda: The Comic Darwin

T his book, which began with wonder at Darwin's remarkable imagination and the quiet power of his writing, should, and will, end with wonder, reproducing as best I can the double-movement pattern that shapes the *Origin* and all the fundamental arguments of that great book. Darwin's writing, full of personal expression of wonder at Nature's remarkable ways, opens up to other worlds, apparently un-Darwinian, in celebration of art, and the dexterity and imagination of human consciousness, while explaining it all in terms of the ordinary. Paradox takes its paradoxical ways among writers distinctly unscientific but responsive to Darwin's new way of understanding the world. In the end, this book will have done what work I could have hoped for if it manages two things: first, to recognize that the words Darwin used count, and that the "meaning" of his work inheres as much in the nuances of feeling, the affirmation of an engaged self, and the texture of his arguments as in the ideas that have done so much to open up the world of organic life, and ourselves; and second, a consequence of the first, to shift for a moment the focus from Darwin, the theorist of endless combat and tragic loss, to Darwin the artist-celebrant of an all-too-imperfect world, which it is far better for us to understand than to accept in the form of still dominant interpretations that require surrender to the notion that the world is either bleakly alien, fit only for tragedy and disenchantment, or somehow in the hands of a transcendent and redeeming power. Wilde worked this out by juxtaposing and opposing art to Darwin's relentless and sometimes cruel nature, with all of its nasty black bugs. Pater took advantage of the Darwinian representation of flux and change to focus attention on what it is that the human mind can do and know, and to gather the extraordinary beauty of particular moments and things. Hardy found ways to make the strenuously mindless operations of nature very beautiful, while breaking through old forms and conventions.

Darwin himself looked upon the world he unfolded for the nineteenth century with an enthusiasm that is rarely attended to in the general haste to applaud or condemn his conception of a mindless world governed always by natural laws that can be understood with no reference to any lawgiver. Although the general view is right that Darwin's world is full of competition and ruthless destruction, reading what he actually wrote tells us a more complicated and ultimately a somewhat different story. And in my conclusion I want to talk about Darwin's "Conclusion," for one would have to be deaf and insensitive not to hear its enthusiasm and excitement, and not to feel the extraordinary pleasure in multiplicity, diversity, contrivance, adaptation, and the simply beautiful. There is nothing here to lament; no loss to mourn. It's as though, now stepping back from the detailed arguments (most, he believed, needing to be filled out yet further by later books), he focuses with a fresh clarity on what he has wrought.

He begins that last chapter as he begins every argument, announcing with something like awe at the difficulty of the enterprise: "Nothing at first can appear more difficult to believe than that the more complex organs and instincts should have been perfected, not by means superior to, though analogous with, human reason, but by the accumulation of innumerable slight variations, each good for the possessor" (p. 459). Trial and error, with no element of human reason, instead of conscious design. From page to page, the "Conclusion" addresses the "difficulty" confidently, and as it does so it points to the future optimistically, and finds energy and possibility everywhere. Although all good biologists know that "perfection," in an absolute sense, is simply impossible in Darwin's world, it is worth noticing that word, "perfected," coming right near the end of his book. There is the experience of perfection, even if part of what we learn is that it is not perfection at all.

Darwin's "Conclusion" is surely comic in the largest sense of that term. There is even a direct polemic energy that for the most part Darwin eschewed through the body of the *Origin*. "By no means," does Darwin "expect to convince experienced naturalists whose minds are stocked with a multitude of facts all viewed, during a long course of years, from a point of view directly opposite to mine." And then, unexpectedly for readers accustomed through almost five-hundred pages to the tone of Darwin's voice, he rather contemptuously comments, "It is so easy to hide our ignorance under such expressions as the 'plan of creation,' 'unity of design,' &c," where the ampersand feels extremely dismissive. It is so easy "to think that we give an explanation when we only restate a fact. Any one whose disposition leads him to attach more weight to unexplained difficulties than to the explanation of a certain number of facts will certainly reject my theory." And so he dismisses those "experienced naturalists," and anticipates critiques, even as they continue today, from those who, finding that scientists are not ready at the moment to explain *everything*, dismiss the science that has done so much extraordinary explanation already. He knows that he faces a "load of prejudice by which the subject is overwhelmed." He knows those prejudices need to "be removed" (p. 482).

His dismissal of objections that follow is also almost derisive, although Darwin never publically loses his temper or stops writing like a gentleman. He mocks the inconsistency of those who on the one hand recognize that some species are not real after all, but believe nevertheless that others *are* real and have been independently created. Although they concede that some species they had thought were real are not, "they refuse to extend the same view to other and very slightly different forms." These authors, he notes in something like puzzled contempt, "seem no more startled at a miraculous act of creation than at an ordinary birth. But do they really believe that at innumerable periods in the earth's history certain elemental atoms have been commanded suddenly to flash into living tissues?" Once again, finally, the burden of proof shifts from the Darwin who has labored through enormous tracts of fact to find adequate explanation to those who propose such an improbable one. Although there have been moments of this sort throughout the book, here they emerge in a tone that is almost triumphant. How absurd the counter-arguments look, he seems to say. How inconsistent are the arguments. Yes, of course, there are things *I* can't explain, but *their* responsibility is to explain them better—and they produce only tautological or inconsistent answers.

But it is not to this somewhat atypical Darwinian prose that I want to turn finally, in representing a Darwin far from the bleak, sad-looking bearded figure he is so often represented to be. Surely that old guy had fun playing the piano for worms and having his children play other instruments for them. Surely he had fun gathering the mud for his experiments at home, bending in the field counting the rings of pines that had year after year been chewed as seedlings by cattle; surely he loved not only his exotic youthful engagement with the tropics, but his walks on the sand path at home, the testing of seeds in salt water, the endless correspondence with naturalists all over the world. Some of the pleasure and enthusiasm of those aspects of Darwin emerge very clearly as he thinks about what all this fun and hard work has produced.

He continues to be all but overwhelmed by his discovery that "all the members of whole classes can be connected together by chains of affinities, and all can be classified on the same principle, in groups subordinate to groups." He has found order in the natural world through all its entanglements and apparent disorder, an order that can, of course, only be explained adequately by his theory of inheritance. It is not that sharp, clean, perfect order that

Natural Theology has implied, but it is an order that allows understanding—if it also induces awe. And it brings him to one of the thrilling ideas in the history of science, and culture, although it is so well understood now as to seem perhaps banal: first, "I believe that animals have descended from at most only four or five progenitors, and plants from an equal or lesser number," and then second, taking the idea to its limit, "Analogy would lead me one step further, namely, to the belief that all animals and plants have descended from some one prototype" (p. 484). Touchingly, after this stunning pronouncement, the modest Darwin re-emerges, the careful Darwin, and he confesses (one can almost hear the sigh), "But analogy may be a deceitful guide." And yet Darwin's world was largely constructed out of analogy, the extraordinary power that allowed him to imagine connections that nobody else had been able to see, that no other mind had been able to make.

There is the combination of a personally present, modest, and careful Darwin, whose voice we have heard throughout the book, with a fiercely committed and almost adamant belief in what he has discovered, in what, throughout the book, he has called "my theory." And so, after agreeing that analogy won't quite do, he nevertheless goes on to provide powerful reasons to think that we are in fact "all descended from one prototype": chemical similarities, for example. And he returns, despite his moderate hesitations, to evoke analogy as evidence once more: "Therefore I would infer from analogy that probably all the organic beings which have ever lived on this earth have descended from some one primordial form, into which life was first breathed" (p. 484).

Despite the inevitable Darwinian hiccup of hesitation—"probably"—it is an astonishing, a breath-taking assertion, hardly occasion for moaning and gnashing of teeth. "The force that through the green fuse drives the flower/drives my green age." There is even spirit, if anonymous spirit here. And as he returns to the idea in this paragraph, it gathers to its greatest force—"all the organic beings which have ever lived on this earth." That means all the subjects he has engaged in detail, the orchids and aphids and ants and worms and wasps and parasites and monkeys and hippopotami, and some that he has not—that is, us. Darwin himself; you, the reader. It is hard not to feel the exhilaration about which Gopnik talks, to feel that the *Origin* is a book "that makes the whole world vibrate."

The very metaphorical Darwin makes yet other optimistic and interesting moves. Most particularly, when he talks about the effect of his ideas on "other departments of natural history," he says that:

The terms used by naturalists of affinity, relationship, community of type, paternity, morphology, adaptive characters, rudimentary and aborted organs, &c., will cease to be metaphorical, and will have a plain signification. When we no longer look at an organic being as a savage looks at a ship, as at something wholly beyond his comprehension; when we regard every production of nature as one which has had a history; when we contemplate every complex structure and instinct as the summing up of many contrivances, each useful to the possessor, nearly in the same way as when we look at any great mechanical invention as the summing up of the labour, the experience, the reason, and even the blunders of numerous workmen; when we thus view each organic being, how far more interesting, I speak from experience, will the study of natural history become! (p. 486)

If this weren't a coda, I would linger yet longer over this passage, take pleasure in its periodic flow, in its growth toward that excited exclamation point, and in its carefully modulated rhythms; I would pause to consider the emotional and intellectual implications of its almost breathless representation of Darwin's own excitement— "I speak from experience"—and note that with all that excitement it provides an excellent and responsible summary of much that this book has been trying to achieve. Returning through the endless details of organic life, Darwin asserts a view normally associated with religion and always understood as metaphorical—we are all one family. But that argument, claims Darwin, "will cease to be metaphorical, and will have a plain signification." As Beer puts it, here Darwin substantiates metaphor, and finds "a real place in the natural order for older mythological expressions" (p. 74). And the act is a thrilling one, the end of a romantic voyage of discovery— "when we thus view each organic being, how far more interesting, I speak from experience, will the study of natural history become!" The exclamation point is Darwin's, not mine. It culminates in Darwin's assertion of his own experience as evidence, and turns all that dogged hard work, all that rummaging in domesticity and ordinariness, into something triumphant and awe-inspiring. He is in effect here asking readers to feel the consequences of his argument as much as to reason toward his conclusion. The implicit narrative here is the growth of humanity into mature and full consciousness, and it sounds rather more like the end of a great and successful adventure story than like the solemn and funereal conclusions of tragedy.

The world represented in Darwin's prose, from the days of the *Beagle* through the arguments of the *Origin* and *Descent*, is full of possibilities and meaning. History is embedded in every moment, consanguinity in every living organism and every extinct one, relationships are implicit everywhere. Analogy can become homology. All of the world is a rastro, Darwinian eyes will trace it, Darwinian exuberance and wonder precede and follow from it. Eyes will trace it by means of that uncanny faculty of complex consciousness, which is itself, in Darwin's view, also a result of the slow combinations of material nature, and which becomes nature's apex. It's not only the watch that Paley finds on his walk, but the stone itself; every production of nature becomes to that consciousness, significant. And although there is a mechanical analogy here, the important thing to note is that it is an analogy that marks a difference from the watch. We have, along the lines of Paley's natural theology, tended to think of organisms on the mechanical model. But Darwin's "designed" and meaningful world does not move on strictly mechanical principles. We know from the rest of the book that "the labour...experience...reason...blunder of numerous workmen" are rather the processes of time under the unchanging laws of nature. The "blunders" are important, a key part of Darwin's legacy to those future thinkers who will come closer to getting it all right. It is part of Darwin's rhetorical strategy and at the center of his thinking that blunders discovered are a key to nature's real design; the mind tracing those blunders, working the analogies, distinguishing the homologies, writing the histories, is the designer.

Wilde understood that and ran with the idea. There are undoubtedly lots of better ways to build whales, and bats, and people's eyes, but the process of trial and error, the chance-driven changes in climate and environment, the "laws" of heredity—all, in Darwin's imagination, account for the strange awkwardness of so much of life, it's refusal to be one thing in one place and perfectly so. The blunders are the most helpful clues to history, and the strongest evidence for Darwin's theory, and essential to the strange marvel of Darwin's nature and his prose.

The "Conclusion," with all these blunders yet in mind, is unequivocally confident, future-directed, and celebratory. Nature makes blunders that the mind discovers, refusing to be deceived by the delicacy and precision of adaptation. It is *no* wonder that writers throughout the late nineteenth century found Darwin's vision and his prose inescapable.

I vowed to myself when I began writing this book that I would steer clear of the *Origin*'s famous last paragraph: everyone knows about it, everyone who writes about Darwin ends up by citing it, and usually as a climax to the argument, since that paragraph is undoubtedly better than anything any student of Darwin could write. But I've given up hiding from it after all; it does too much work to ignore, and represents too richly the complicated beauty of Darwin's argument and his art. I need it, in particular, because I want so strongly to emphasize what is in a sense the overall argument of this book, that latent and often overt in Darwin's prose is a Darwin very different from the tragic Darwin. This other mindful Darwin emerges in the art of his prose, often in tension with its most overt statements about the way the world is. The tree of life is beautiful and fascinating but its power as metaphor comes partly from the way it turns the inevitable deaths and chancy cruelty of nature into the loveliness of a real tree on which branches produce only twigs that die while others ramify extraordinarily. It is a tree that looks like many we have all experienced. But under the pressure of Darwin's imagination it is transformed so that its very shape bears a significance that its apparently chancy form would not otherwise intimate. The imagination is not sentimental or unequivocally upbeat—part of its force lies in its recognition of the number of dead twigs and unproductive branches. But the tree is exuberantly alive.

The third move in the characteristic pattern of Darwin's prose, revealing that the ordinary means by which the difficult even "staggering" phenomenon has come to be are themselves wonderful, is virtually the same move that realist novelists make when, as George Eliot put it in her famous manifesto (chapter seventeen) of *Adam Bede*, the same year as the *Origin* was published: "I turn without shrinking from cloud-borne angels, from prophets, sibyls, and heroic warriors to an old woman bending over her flower pot."[1] The old woman over the flower pot, Darwin over his garden weeds in a cup: they are both important and yet both within the range of ordinary understanding. They are both deeply moving. Isn't it amazing that such strange phenomena are really ordinary after all, manifestations of ordinary causes? In spite of everything, Darwin is enjoying the paradox of life after death, the paradox of design undesigned, the paradox of mindlessness producing the mind that can detect it, and most of all, I like to believe, Darwin is enjoying himself (as we should be doing too). As Culler insists, other writers, even some who otherwise took the bleak message of natural selection quite unmetaphorically, understood that this was so. And modernist literature, whose central trope is paradox, would be marked as much but not so obviously by Darwin's comedy as by his tragedy.

The voice of the genial, modest but adamant, counter-intuitive, imaginatively daring, and scrupulously empirical scientist is nowhere more forcefully present than in that last paragraph, where he recapitulates the book's very form, and he can at the same time just relax into writing well, or perhaps better, make one more great metaphorical leap to a summation, literal and Romantic, of what he had wrought. The summary is brilliantly succinct, taking us in a few lines through the whole argument, from fascination, to the laws of nature, to grandeur, and, representing nature by a familiar scene and making us aware of its complexities and wonders, teaching us that every worm has its history, every tangled bush is a rastro. The nature revealed in the tangled bank is here so clearly and splendidly articulated that its counter-intuitive nature seems to have become intuitive. The hitherto astonishing counter-intuitive world finds a beautiful form, the complications of feeling and move-ment emerge "entangled," the direct confrontation with the worst emerges in a representation of the beauty and possibilities of a world that is also frightening, both complex and simple, in which every-thing depends on everything else, and all things are related. As we look at that famous last paragraph again, I urge only some more careful attention to all the remarkable work it does, from word to word, sentence to sentence. The old double movement is reaffirmed and embodied, as this intricately entangled world is explained ac-cording to the laws that produced it, and yet, as each detail in the scene gets the attention and respect Darwin gives to all of nature, as the assumptions about the way the world works are made explicit, and the extraordinary diversity and richness of all that highly designed life with all that history are intimated, in the end that nature and that prose emerge in all their ordinariness, quite wonderful again.

It is interesting to contemplate an entangled bank, clothed with many plants of many kinds, with birds singing on the bushes, with various insects flitting about, and with worms crawling through the damp earth, and to reflect that these elaborately constructed forms, so different from each other and depen-dent on each other in so complex a manner, have all been produced by laws acting around us. These laws, taken in the largest sense, being Growth with

Reproduction; Inheritance which is almost implied by reproduction; Variability from the indirect and direct action of the external conditions of life, and from use and disuse; a Ratio of Increase so high as to lead to a Struggle of Life, and as a consequence to Natural Selection, entailing Divergence of Character and the Extinction of less-improved forms. Thus, from the war of nature, from famine and death, the most exalted object which we are capable of conceiving, namely, the production of the higher animals, directly follows. There is grandeur in this view of life, with its several powers, having been originally breathed into a few forms or into one; and that, while this planet has gone cycling on according to the fixed law of gravity, from so simple a beginning endless forms most beautiful and most wonderful have been, and are being evolved. (p. 490)

Note

1. George Eliot, *Adam Bede* (New York: Oxford University Press, 1996; 1859), 177.

SELECT BIBLIOGRAPHY ON DARWIN, HIS THEORY, AND ITS EXTRASCIENTIFIC CONNECTIONS

Amigoni, David, *Colonies, Cults and Evolution: Literature, Science and Culture in* Nineteenth Century Writing (Cambridge: Cambridge University Press, 2007).

—— and Wallace, Jeff (eds.), *Charles Darwin's "The Origin of Species"* (Manchester: Manchester University Press, 1995).

——, Barlow, Paul, and Trodd Colin (eds.), Victorian Culture and the Idea of the Grotesque (Aldeshot: Ashgate, 1999).

Appleman, Philip, *Darwin: A Norton Critical Edition*, 3rd edn. (New York: W. W. Norton, 2010).

Barrett, Paul H., Weinshank, Donald, and Gottleber, Timothy, *A Concordance to Darwin's "Origin of Species,"* 1st edn (Ithaca: Cornell University Press, 1981).

——, Gautrey, P. J., Herbert, S., Kohn, D., and Smith, S. (eds.), *Charles Darwin's Notebooks, 1836–1844* (Cambridge: Cambridge University Press, 1987).

Barrish, Philip, "Accumulating Variation: Darwin's *On the Origin of Species* and Contemporary Literary and Cultural Theory," *Victorian Studies* 34 (1991): 431–54.

Bayley, John, *The Realistic Imaginations* (Chicago: University of Chicago Press, 1981).

Beatty, John, "What's in a Word? Coming to Terms in the Darwinian Revolution," *Journal of the History of Biology* 15 (1982): 215–39.

Beer, Gillian, *Open Fields: Science in Cultural Encounter* (Oxford: Oxford University Press, 1996).

—— *Darwin's Plots: Evolutionary Narrative in Darwin, George Eliot and Nineteenth-Century Fiction*, 2nd edn. (London: Routledge, 1983); 3rd edn (Cambridge University Press, 2009).

Bowlby, John, *Charles Darwin: A New Life* (New York: W. W. Norton, 1991).

Bowler, Peter J., *Evolution: The History of an Idea* (Berkeley: University of California Press, 1984).

—— *The Non-Darwinian Revolution: Reinterpreting a Historical Myth* (Baltimore: Johns Hopkins University Press, 1988).

—— *The Invention of Progress* (Oxford: Basil Blackwell, 1989).

Bowler, Peter J., *Charles Darwin: The Man and His Influence* (Oxford: Basil Blackwell, 1990).

Bronowski, Jacob, *The Common Sense of Science* (Cambridge, MA: Harvard University Press, 1978).

Boyd, Brian, *The Origins of Knowledge and Imagination* (New Haven: Yale University Press, 1978).

—— *On the Origin of Stories: Evolution, Cognition, and Fiction* (Cambridge, MA: Harvard University Press, 2009).

Brown, Andrew, *The Darwin Wars: How Stupid Genes Became Selfish Gods* (New York: Simon and Schuster, 1999).

Browne, Janet, *Voyaging: Charles Darwin: A Biography* (New York: Knopf, 1995).

—— *Charles Darwin: The Origin and After—The Years of Fame* (New York: Knopf, 2002).

—— (eds.), *The Correspondence of Charles Darwin (CCD)* Vol.9. (Cambridge University Press, 1994).

Burckhardt, Frederick *et al.* (eds.), *The Correspondence of Charles Darwin (CCD)*, Vol.6. (Cambridge: Cambridge University Press, 1996).

—— (eds.), *The Correspondence of Charles Darwin (CCD)* Vol. 9 (Cambridge: Cambridge University Press, 1994).

Butler, Samuel, *Life and Habit* (London: A. C. Fifield, 1878; 1910).

Carroll, Joseph, *Evolution and Literary Theory* (Columbia: University of Missouri Press, 1995).

Caudill, Edward, *Darwinian Myths: The Legends and Misuses of a Theory* (Knoxville: University of Tennessee Press, 1997).

Chambers, Robert, *Vestiges of the Natural History of Creation* (1844; rept, Leicester University Press, 1969).

Chapple, J. A. V., *Science and Literature in the Nineteenth Century* (London: MacMillan, 1986).

Churchill, Frederick, "Darwin and the Historians," in *Charles Darwin. A Commemoration, 1882-1892*, ed. R. J. Berry (London: Academic Press, 1982).

Clark, Ronald W., *The Survival of Charles Darwin: A Biography of a Man and an Idea* (New York: Random House, 1984).

Clifford, W. K., "The Philosophy of the Pure Sciences," *Lectures and Essays*, eds. Leslie Stephen and Sir Richard Pollock (London: MacMillan & Co., 1901).

Colley, Ann C., "Nostalgia in *The Voyage of the Beagle*," *Centennial Review* 35 (Winter, 1991): 167–83.

Colp, Ralph Jr, "'I Never Wrote So Much About Myself': Charles Darwin's 1861–1870 Autobiographical Notes," in *Darwin Today: The 8th Kuhlungsborn Colloquium on Philosophical and Ethical Problems of Biosciences*, eds. E. Geissler and W. Scheler, Abhandlungen der Akademie der Wissenschafter der DDD, Abteilung Mathematick—Naturwissenschaften Technik, 1983 (Berlin: Akademie Verlag, 1983), 37–51.

—— "Notes on Charles Darwin's *Autobiography*," *Journal of the History of Biology*, 18 (Fall 1985), 357–401.

—— "'Confessing a Murder': Darwin's First Revelations about Transmutation," *Isis* 77 (1986): 9–32.

Cook, E. T. and Wedderbarn, A. (eds.), *The Works of John Ruskin* (London: George Allen, 1904).

Cox, R.G. (ed.), *Thomas Hardy: The Critical Heritage* (London: Routledge & Kegan Paul, 1970), 142.

Cronin, Helen, *The Ant and the Peacock* (Cambridge: Cambridge University Press, 1991).

Crook, Paul, *Darwinism, War, and History* (Cambridge: Cambridge University Press, 1994).

Culler, A. Dwight, "The Darwinian Revolution and Literary Form," in George Levine and William Madden, *The Art of Victorian Prose* (New York: Oxford University Press, 1968).

Dale, Peter Alan, *In Pursuit of a Scientific Culture: Science, Art, and Society in the Victorian Age* (Madison: University of Wisconsin Press, 1989).

Darwin, Charles, *The Voyage of the Beagle*, ed. Leonard Engel (New York: The Natural History Library, Anchor Books, [1839] 1962); 500.

Darwin, Charles and Wallace, A. R., *Evolution by Natural Selection*, ed. Gavin de Beer (Cambridge: Cambridge University Press, 1958) (contains Darwin's 1842 and 1844 drafts of his theory, and Wallace's 1858 paper that precipitated the writing of the *Origin*, also listed below).

Darwin, Francis (ed.), The Foundations of the Origin of Species (BiblioBazaar, [1909] 2008), 86.

Dawkins, Richard, *River Out of Eden* (New York: Basic Books, 1995).

—— *Unweaving the Rainbow: Science, Delusion and the Appetite for Wonder* (New York: Houghton Mifflin, 1998).

—— *A Devil's Chaplain: Reflections on Hope, Science, and Love* (New York: Houghton Mifflin, 2003).

—— *The Greatest Show on Earth: The Evidence for Evolution* (New York: Free Press, 2009), 408–13.

Dawson, Gowan, *Darwin, Literature and Victorian Respectability* (Cambridge: Cambridge University Press, 2007).

Dennett, Daniel C., *Darwin's Dangerous Idea: Evolution and the Meaning of Life* (New York: Simon and Schuster, 1995).

Depew, David J., "The Rhetoric of the *Origin of Species*," in *The Cambridge Companion to the "Origin of Species"*, eds. Michael Ruse and Robert J. Richards (Cambridge: Cambridge University Press, 2009).

Desmond, Adrian, *Archetypes and Ancestors: Palaeontology in Victorian London, 1850–1875* (Chicago: University of Chicago Press, 1984).

—— *The Politics of Evolution: Morphology, Medecine, and Reform in Radical London* (Chicago: University of Chicago Press, 1989).

—— and James Moore, *Charles Darwin: The Life of a Tormented Evolutionist* (London: Michael Joseph, 1991).

—— —— *Darwin's Sacred Cause: Race, Slavery and the Quest for Human Origins* (New York: Houghton Mifflin, 2009).

Dewey, John, *The Influence of Darwin on Philosophy* (New York: Holt, 1910).

Donald, Diana and Muntro, Jane, *Endless Forms: Charles Darwin, Natural Science and the Visual Arts* (New Haven: Yale University Press, 2009).

Doyle, Arthur Conan, *Through the Magic Door* (Pleasantville: Akadine Press, 1999; 1907).

—— *Sherlock Holmes: The Complete Novels and Stories* (New York: Bantam Books, 1986) I, 14.

Durant, John *Darwin and Divinity* (Oxford: Blackwell, 1985).

Ebbatson, Roger, *The Evolutionary Self: Hardy, Forster, Lawrence* (Brighton: Harvester Press, 1982).

Eiseley, Loren, *Darwin's Century* (Garden City: Doubleday, [1958], 1961).

Eldredge, Niles, *Darwin: Discovering the Tree of Life* (New York: W. W. Norton, 2005).

Eliot George, *Adam Bede* (New York: Oxford University Press, [1859] 1996), 177.

—— *Middlemarch: A Study of Provincial Life* (Oxford: Oxford University Press, 1996), 162.

Ellegard, Allvar, *Darwin and the General Reader: The Reception of Darwin's Theory of Evolution in the British Periodical Press, 1859–1872* (Chicago: University of Chicago Press, [1958] 1990).

Elleman, Richard (ed.), *Critical Writings of Oscar Wilde (New York: Random House, 1968), 291.*

Fichman, Martin, *Evolutionary Theory and Victorian Culture* (Amherst, NY: Humanity Books, 2002).

Gale, Barry, "Darwin and the Concept of a Struggle for Existence: A Study of the Extrascientific Origins of Scientific Ideas," *Isis* 63 (1972): 321–44.

—— *Evolution Without Evidence* (Brighton: Harvester, 1982).

Gaul, Marilyn, "From Wordsworth to Darwin: 'On to the Fields of Praise,'" *The Wordsworth Circle* 10(1), (Winter 1979). (Boston, MA: Boston University. Editorial Institute, 1979).

Gerth, H. H. and Wright Mills, C. (eds.), *From Max Weber* (Oxford: Oxford University Press, 1958, 1946), 129–58.

Ghiselin, Michael, *The Triumph of the Darwinian Method* (Berkeley: University of California Press, 1969).

Gillispie, C. C., *Genesis and Geology* (Cambridge: Cambridge University Press, [1951] 1959).

Gillespie, Neal C., *Charles Darwin and the Problem of Creation* (Chicago: University of Chicago Press, 1979).

Gould, Stephen Jay, *Ever Since Darwin* (New York: W. W. Norton, 1977a).

—— *Ontogeny and Phylogeny* (Cambridge: Harvard University Press, 1977b).

—— *The Panda's Thumb* (New York: W. W. Norton, 1982).

—— *Time's Arrow, Time's Cycle* (Cambridge: Harvard University Press, 1987).

Gopnik, Adam, *Angels and Ages: A Short Book about Darwin, Lincoln, and Modern Life* (New York: Knopf, 2009).

Grant, K. Thalia and Estes, Gregory B., *Darwin in Galapagos: Footsteps to a New World* (Princeton: Princeton University Press, 2009).

Greene, John C., *The Death of Adam: Evolution and its Impact on Western Thought* (New York: Mentor, 1959).

—— *Science, Ideology, and World View: Essays in the History of Evolutionary Ideas* (Berkeley: University of California Press, 1981).

—— *Debating Darwin* (Claremont: Regina Press, 1999).

Gruber, Howard, *Darwin on Man* (London: Dutton, 1974).

Hardy, Thomas, *Return of the Native* (London: Penguin Books, 1983).

—— *Tess of the D'Urbervilles* (Oxford: University Press, 1988), Part 5, Ch. 35, p. 225.

—— *The Woodlanders* (London: Penguin Books, 1998), 22.

Herbert, Sandra, "The Place of Man in the Development of Darwin's Theory of Transmutation," *Journal of the History of Biology* 10 (1977): 155–227.

Hodge, Jonathan and Radick, Gregory, *The Cambridge Companion to Darwin* (Cambridge: Cambridge University Press, 2003).

Holloway, John, *The Victorian Sage: Studies in Argument* (New York: W. W Norton, 1953), 9–10.

Holmes, John, *Darwin's Bards: British and American Poetry in the Age of Evolution* (Edinburgh: Edinburgh University Press, 2009): xiv, 288.

Holmes, Richard, *The Age of Wonder: How the Romantic Generation Discovered the Beauty and Terror of Science* (New York: Pantheon Books, 2009).

Hoslë, Vittorio and Hilles, Christian, *Darwinism & Philosophy* (South Bend: University of Notre Dame Press, 2005).

Hull, David L., *Darwin and his Critics: The Reception of Darwin's Theory of Evolution by the Scientific Community* (Chicago: University of Chicago Press, 1973).

—— *Man's Place in Nature and Other Anthropological Essays* (London: Macmillan, [1863], 1894).

—— *Methods and Results* (London: MacMillan, 1893).

—— *Darwiniana* (London: MacMillan, 1897).

Huxley, T. H., *Life and Letters*, ed. Leonard Huxley, 2 Vols (London: MacMillan, 1900).

Hyman, Stanley Edgar, *The Tangled Bank: Darwin, Marx, Frazer and Freud as Imaginative Writers* (New York: Atheneum, 1962).

Jann, Rosemary, "Darwin and the Anthropologists: Sexual Selection and its Discontents," *Victorian Studies* 37 (1994): 286–306.

Jehlen, Myra, *Five Fictions in Search of Truth* (Princeton: Princeton University Press, 2008).

Jehlen, Myra, "On How to Become Knowledge, Cognition Needs Beauty," *Raritan* 39: September 2010, 39–46.

Jones, Greta, *Social Darwinism and English Thought: The Interaction Between Biological and Social Theory* (Brighton: Harvester Press, 1980).

Jones, Steve, *Darwin's Ghost: The Origin of Species Updated* (New York: Ballantine Books, 2000).

Keynes, Randal (ed.), *Charles Darwin's "Beagle" Diary* (Cambridge: Cambridge University Press, 1988).

—— *Darwin, His Daughter and Human Evolution* (New York: Riverhead Books, 2002a).

—— *Fossils, Finches and Fuegians* (London: Harper Collins, 2002b).

—— *Living with Darwin: Evolution, Design, and the Future of Faith* (Oxford: Oxford University Press, 2007).

Keynes, Richard (ed.), *Charles Darwine's Zoology Notes and Specimen Lists from H.M.S Beagle* (Cambridge: Cambridge University Press, 2000).

Kitcher, Philip, *The Achievement of Science: Science without Legend, Objectivity without Illusions* (Oxford: Oxford University Press, 1993) (chapter 2, on Darwin).

Knight, David, *The Age of Science* (Oxford: Blackwell, 1986).

Kohn, David, "Theories to Work By: Rejected Theories, Reproduction, and Darwin's Path to Natural Selection," *Studies in History of Biology* (1980) 4: 67–170.

—— (ed.), *The Darwinian Heritage* (Princeton: Princeton University Press, 1985)

—— "Darwin's Ambiguity: The Secularization of Biological Meaning," *British Journal for the History of Science* 22 (1989): 215–39.

—— "The Aesthetic Construction of Darwin's Theory," in Alfred I. Tauber (ed.), *The Elusive Synthesis: Aesthetics and Science* (Dordrecht: Kluwer, 1996): 48.

Knoepflmacher, U. C. and Tennyson, G. B. (eds.), *Nature and the Victorian Imagination* (Princeton: Princeton University Press, 1977) (See essays by Roger Smith, "The Human Significance of Biology: Carpenter, Darwin, and the *vera causa*"; and David B. Wilson, "Concepts of Physical Nature: John Herschel to Karl Pearson".)

Kramer, Dale, *Critical Essays on Thomas Hardy: The Novels* (Boston: G. K. Hall 1990), 191.

Krasner, James, "A Chaos of Delight: Perception and Illusion in Darwin's Scientific Writing," *Representations*" (1990) 31: 118–41.

—— *The Entangled Eye: Visual Perception and the Representation of Nature in Post-Darwinian Narrative* (Oxford: Oxford University Press, 1992).

Kropotkin, Peter, *Mutual Aid: A Factor of Evolution* (New York: McClure Phillips & Co., 1902).

Lamarck, J. B., *Zoological Philosophy: An Exposition with Regard to the Natural History of Animals* (Chicago: University of Chicago Press, 1984).

Levine, George and Madden, William A. (eds.), *The Art of Victorian Prose* (New York: Oxford University Press, 1968), 225.

—— "Darwin and the Problem of Authority," *Raritan Review* 3 (Spring, 1984): 30–63.

—— (ed.), *One Culture: Essays in Science and Literature* (Madison: University of Wisconsin Press, 1987).

—— *Darwin and the Novelists* (Cambridge, MA: Harvard University Press, 1988).

—— "Charles Darwin's Reluctant Revolution," *SAQ*, 91 (Summer, 1992): 525–56.

—— "By Knowledge Possessed: Darwin, Nature, and Victorian Narrative," *New Literary History* 24 (Spring, 1993): 363–92.

—— "Darwin and Pain: Why Science Made Shakespeare Nauseating," *Raritan Quarterly* 15 (Fall, 1995): 97–144.

—— *Darwin Loves You: Natural Selection and the Re-enchantment of the World* (Princeton: Princeton University Press, 2006).

—— "Ruskin and Darwin and the Matter of Matter," in *Realism, Ethics, and Secularism: Essays on Victorian Science and Literature* (Cambridge: Cambridge University Press, 2008).

—— (ed.), *The Joy of Secularism: Eleven Essays on How We Live Now* (Princeton: Princeton University Press, 2011).

Lewes, G. H., *Studies in Animal Life* (London: Smith, Elder, and Co., 1862).

Lyell, Charles, *Principles of Geology* (Chicago: University of Chicago Press, [1830] 1990).

Lightman, Bernard, *The Origins of Agnosticism: Victorian Unbelief and the Limits of Knowledge* (Baltimore: The Johns Hopkins University Press, 1987).

Mandelbaum, Maurice, *History, Man, and Reason: A Study in Nineteenth-Century Thought* (Baltimore: Johns Hopkins University Press, 1971).

Manier, Edward, *The Young Darwin and his Cultural Circle* (South Bend: University of Notre Dame, 1978).

Mayr, Ernst, *Evolution and the Diversity of Life* (Cambridge: Harvard University, 1976).

—— *The Growth of Biological Thought* (Cambridge: Harvard University Press, 1982).

—— *One Long Argument: Charles Darwin and the Genesis of Modern Evolutionary Thought* (Cambridge, MA: Harvard University Press, 1991).

Meisel, Perry, *Thomas Hardy: The Return of the Repressed, A Study of the Major Fiction* (New Haven: Yale University Press, 1972).

Miller, Jonathan, *Darwin for Beginners* (New York: Pantheon, 1982).

Miller, Kenneth R., *Finding Darwin's God: A Scientist's Search for Common Ground Between God and Evolution* (New York: Harper Perennial, 2007).

—— *Only a Theory: Evolution and the Battle for the American Soul* (New York: Viking Adult, 2008).

Millgate, Michael, *Thomas Hardy: His Career as a Novelist* (New York: Random House, 1971).

——(ed.), *The Life and Works of Thomas Hardy* (Athens, GA: University of Georgia Press 1985).

——(ed.), *Thomas Hardy: Selected Letters* (Oxford: Clarendon Press, 1990).

Moore, James R., *The Post-Darwinian Controversies: A Study of the Protestant Struggle to Come to Terms with Darwin in Great Britain and America, 1870-1900* (Cambridge: Cambridge University Press, 1979).

—— "1859 and All That: Remaking the Story of Evolution and Religion," in *Charles Darwin, 1809-1882. A Centennial Commemorative* (Wellington, N. Z.: Nova Pacifica, 1983).

—— (ed.), *History, Humanity and Evolution* (Cambridge: Cambridge University Press, 1989).

—— "Why Darwin Gave up Christianity," in *History, Humanity and Evolution* (Cambridge: Cambridge University Press, 1989).

—— *The Darwin Legend* (Grand Rapids: Baker Books, 1994).

—— "Darwin of Down: The Evolutionist as Squarson-Naturalist," in David Kohn (ed.), *The Darwinian Heritage*.

Morton, Peter, *The Vital Science: Biology and the Literary Imagination: 1860–1900* (London: Allen and Unwin).

Oldroyd, D. R., *Darwinian Impacts: An Introduction to the Darwinian Revolution* (Atlantic Highlands: Humanities Press, 1980).

—— and Ian Langham, (eds.), *The Wider Domain of Evolutionary Thought* (Dordrecht: D. Reidel, 1983) (Eveleen Richards, Rosaleen Love, Michael Ruse).

Ospovat Dov, *The Development of Darwin's* Theory: *Natural History, Natural Theology and Natural Selection, 1838–1859* (Cambridge: Cambridge University Press, 1981).

Paley, William, *Natural Theology, or Evidences of the Existence and Attributes of the Deity, collected from the Appearances of Nature* (Oxford: Oxford University Press, 2008).

Pater, Walter, *The Renaissance: Studies in Art and Poetry* (London: MacMillan and Co., [1873] 1888).

—— "Coleridge." *Appreciations: With an Essay on Tile* (London: MacMillan and Co., [1889] 1922).

—— *Plato and Platonism* (London: MacMillan and Co., 1907), 19–21.

Phillips, Adam, *Darwin's Worms* (New York: Basic Books, 1999).

Postlethwaite, Diana, *Making it Whole: A Victorian Circle and the Shape of Their World* (Columbus: Ohio State University Press, 1984).

Powell, Baden, *Essays on the Spirit of the Inductive Philosophy, the Unity of Worlds, and the Philosophy of Creation* (London: Longman, 1855).

Quammen, David, *The Reluctant Mr Darwin: An Intimate Portrait of Charles Darwin and the Making of His Theory* (New York: W. W. Norton, 2006).

Rachels, James, *Created from Animals: The Moral Implications of Darwinism* (New York: Oxford University Press, 1990).

Richards, Janet Radcliffe, *Human Nature and Darwin* (London: Routledge, 2000).

Richards, Robert J., *Darwin and the Emergence of Evolutionary Theories of Mind and Behaviour* (Chicago: University of Chicago Press, 1987).

—— *The Meaning of Evolution: The Morphological Construction and Ideological Reconstruction of Darwin's Theory* (Chicago: University of Chicago Press, 1992).

—— *The Romantic Conception of Life: Science and Philosophy in the Age of Goethe* (Chicago: University of Chicago Press, 2002).

—— "Biology," in *From Natural Philosophy to the Sciences* (ed.) David Cahan (Chicago: University of Chicago Press, 2003) (extremely helpful bibliographical survey).

—— "Darwinian Enchantment," in *The Joy of Secularism,* ed. George Levine (Princeton: Princeton University Press, 2011) .

Rudwick, M. J. S., *The Meaning of Fossils: Episodes in the History of Palaeontology* (Cambridge: Cambridge University Press, 1972).

—— *The Great Devonian Controversy* (Chicago: University of Chicago Press, 1987).

Ruse, Michael, *The Darwinian Revolution* (Chicago: University of Chicago Press, 1979).

—— *Taking Darwin Seriously* (Amherst, NY: Prometheus Books, 1998).

—— *Darwinism and its Discontents* (Cambridge: Cambridge University Press, 2006).

Ruse, Michael and Richards, Robert J. (eds.), *The Cambridge Companion to the* "Origin of Species" (Cambridge: Cambridge University Press, 2009).

Rylance, Rick, *Victorian Psychology and British Culture: 1850–1880* (Oxford: Oxford University Press, 2000).

Schmitt, Cannon, *Darwin and the Memory of the Human: Evolution, Savages, and South America* (Cambridge: Cambridge University Press, 2009).

Schwartz, Joel S., "Darwin, Wallace, and Huxley, and Vestiges of the Natural History of Creation," *Journal of the History of Biology* 23 (1990): 127–53.

Schweber, Sylvan, S., "The Young Darwin," *Journal of the History of Biology* 12 (1979).

—— "Darwin and the Political Economists: Divergence of Character," *Journal of the History of Biology* 13 (1980a): 195–289.

—— "The Origin of the *Origin Revisited*," *Journal of the History of Biology* 10 (1980b): 223–316.

Secord, James A., *Controversy in Victorian Geology: The Cambrian-Silurian Dispute* (Princeton: Princeton University Press, 1986).

—— *Victorian Sensation: The Extraordinary Publication, Reception, and Secret Authorship of* Vestiges *of the Natural History of Creation* (Chicago: University of Chicago Press, 2000).

Shanahan, Timothy, *The Evolution of Darwinism* (Cambridge: Cambridge University Press, 2004).

Singer, Peter, *Politics, Evolution: A Darwinian Left* (New Haven: Yale University Press, 1999).

Smith, Jonathan, *Seeing Things: Charles Darwin and Victorian Visual Culture* (Cambridge: Cambridge University Press, 2007).

Smith II, Philip E. and Helfand, Michael (eds.), *Oscar Wilde's Oxford Notebooks* (New York: Oxford University Press, 1989), 43–5.

Stevenson, Lionel, Thomas, Ronald, "where collins and the Sensation Novel," in John Richeth (ed.) the Columbia history of the British Novel (New york: Columbia University Press, 1994), 506. *Darwin Among the Poets* (Chicago: University of Chicago Press, 1932).

Stott, Rebecca, *Darwin and the Barnacle* (London: Faber and Faber, 2003).

Sulloway, Frank, "Darwin and his Finches: The Evolution of a Legend," *Journal of the History of Biology* 15 (1982a): 1–53.

—— "Darwin's Conversion: the *Beagle* Voyage and its Aftermath," *Journal of the History of Biology* 15 (1982b): 325–96.

Thomas, Ronald, "Wilkie Collins and the Sensation Novel," in John Richetti (ed.), *The Columbia History of the British Novel* (New York: Columbia University Press, 1994), 506.

Thomson, Keith, *Before Darwin: Reconciling God and Nature* (New Haven: Yale University Press, 2005).

Toulmin, Stephen and Goodfield, June, *The Discovery of Time* (Chicago: University of Chicago Press, 1965).

Wallace, Alfred Russel, "On the Tendency of Varieties to Depart Indefinitely from the Original Type," (1858) (Published with Darwin's paper, and notes by Hooker and Lyell in *Journal of the Proceedings of the Linnaean Society: Zoology* 3 [1859]: 45–62).

—— *Darwinism* (London: MacMillan, 1912).

Whewell, William, *Selected Writings on the History of Science*, ed. Yehuda Elkana (Chicago: University of Chicago Press, 1984).

Whitehead, Alfred North, *Science and the Modern World* (New York, Mentor, [1925] 1948).

Wilde, Oscar, *Critical Writings of Oscar Wilde* ed. Richard Ellman (New York: Random House, 1968).

Willey, Basil, *Darwin and Butler: Two Versions of Evolution* (London: Chatto & Windus, 1960).

Willingham-McClain, Gary, "Darwin's 'Eye of Reason': Natural Selection and the Mathematical Sublime," *Victorian Literature and Culture* 25 (1997): 67–86.

Young, Robert, *Darwin's Metaphor: Nature's Place in Victorian Culture* (Cambridge: Cambridge University Press, 1985).

—— "Darwin and the Genre of Biography," in *One Culture: Essays in Science and Literature* (Madison: University of Wisconsin Press, 1987).

Wilson, David Sloan, *Evolution for Everyone* (New York: Delacorte Press, 2007).

—— and Gottschall, Jonathan, *The Literary Animal* (Evanston: Northwestern University Press, 2005).

Woolf, Virginia, *The Mark on the Wall* (Richmond: Hogarth Press, 1919).

Selected Works by Darwin

Darwin's *Correspondence* is being published by Cambridge University Press. Nineteen volumes have now appeared, edited by, among others, Frederick Burkhardt and Sydney Smith and most recently, James Secord; they have also produced a *Calendar of the Correspondence of Charles Darwin* (New York & London: Garland, 1985), providing a précis of every letter to appear in the *Correspondence*. The letters are all available and searchable at a website of the Darwin correspondence project: http://www.darwinproject.ac.uk/home.

Equally, there is now also available *Charles Darwin's Notebooks: 1836–1844*, edited by Paul Barrett, Peter Gautrey, Sandra Herbert, and David Kohn (Ithaca: Cornell University Press, 1987).

Richard Keynes has edited *Charles Darwin's Zoology Notes & Specimen Lists from H. M. S. Beagle* (Cambridge: Cambridge University Press, 2000); Alison Pearn has edited *A Voyage Around the World: Charles Darwin and the Beagle Collections in the University of Cambridge* (Cambridge: Cambridge University Press, 2009). Janet Browne and Michael Neve have edited Charles Darwin's *Voyage of the Beagle* (London: Penguin Books).

There is now a website devoted to "The Complete Works of Charles Darwin Online," and includes usable copies of his complete publications, his private papers and mss, along with a ms catalogue, an international bibliography, and virtually anything else Darwinian that you can think of: http://Darwin-online.org.uk/

An edition of Darwin's unpublished work in preparation for the *Origin* is also available: R. C. Stauffer (ed.), *Charles Darwin's Natural Selection: Being the Second Part of his Big Species Book Written From 1856 to 1858* (Cambridge: Cambridge University Press, 1975).

Among Darwin' important books:

Charles Darwin's Beagle Diary, ed. Richard Darwin Keynes (Cambridge: Cambridge University Press, 1988).

Journal of Researches into the Geology and Natural History of the Various Countries Visited by H. M. S. Beagle (1839). Revisions continue until 1845. There are six editions of *On the Origin of Species*, originally published in 1859, the sixth being the usual reprinted one, but we will be using the first edition. A facsimile edition by Harvard University Press (1964) is available.

On the Various Contrivances by which British and Foreign Orchids are Fertilised by Insects, and on the Good Effects of Intercrossing (1862).

The Descent of Man and Selection in Relation to Sex (1871).

The Variation of Animals and Plants Under Domestication, 2 vols (1868 and 1871).

The Expression of the Emotions in Man and Animals (1872).

The Formation of Vegetable Mould through the Action of Worms with Observations of their Habitats (1883).

Metaphysics, Materalism, and the Evolution of Mind, trans. and anno. Paul H. Barrett (Chicago: University of Chicago Press, 1974).

Charles Darwin's Natural Selection, ed. R. C. Stauffer (Cambridge: Cambridge University Press, 1975). (the Unpublished parts of Darwin's "big book").

The Autobiography of Charles Darwin, ed. Nora Barlow (1969).

INDEX